First International Conference of the OIE Reference Laboratories and Collaborating Centres

Florianopolis, Brazil
3-5 December 2006

Volume editors

Michel Lombard
International Association for Biologicals (IABS)

Betty Dodet
Dodet Bioscience, Lyon, France

In collaboration with the Scientific and Technical Department of the OIE

Proceedings of an international conference organised by the
World Organisation for Animal Health – OIE
and the Federal Ministry of Agriculture, Brazil

5 figures and 5 tables, 2007

KARGER

Basel · Freiburg · Paris · London · New York
Bangalore · Bangkok · Singapore . Tokyo · Sydney

Developments in Biologicals

Library of Congress Cataloging-in-Publication Data

International Conference of the OIE Reference Laboratories and Collaborating Centres (1st : 2006 : Florianopolis, Brazil)
First International Conference of the OIE Reference Laboratories and Collaborating Centres, Florianopolis, Brazil, 3-5 December 2006 / volume editors, Michel Lombard, Betty Dodet ; in collaboration with the Scientific and Technical Department of the OIE ; proceedings of an International Conference organised by the World Organisation for Animal Health- OIE and the Federal Ministry of Agriculture, Brazil.
p. ; cm. -- (Developments in biologicals ; v. 128)
Includes bibliographical references.
ISBN-13: 978-3-8055-8321-3 (soft cover : alk. paper)
1. International Office of Epizootics. Reference Laboratories--Congresses.
2. International Office of Epizootics. Collaborating Centres--Congresses.
3. Animals--Diseases--Congresses. I. Lombard, Michel. II. Dodet, Betty.
III. International Office of Epizootics. IV. Brazil. Ministério da Agricultura. V. Title. VI. Series.
[DNLM: 1. International Office of Epizootics. Reference Laboratories. 2. International Office of Epizootics. Collaborating Centres. 3. Animal Diseases--prevention & control--Congresses. 4. Laboratories--Congresses. 5. Communicable Diseases--veterinary--Congresses. 6. International Agencies--Congresses. 7. Public Policy--Congresses. W1 DE997VKF v.128 2007 / SF 740 I61f 2007]
SF605.I57 2006
636.089'6--dc22

2007019942

Bibliographic Indices. This publication is listed in bibliographic services, including Current Contents® and Index Medicus.

Drug dosage. The authors and the publisher have exerted every effort to ensure that drug selection and dosage set forth in this text are in accord with current recommendations and practice at the time of publication. However, in view of ongoing research, changes in government regulations, and the constant flow of information relating to drug therapy and drug reactions, the reader is urged to check the package insert for each drug for any change in indications and dosage and for added warnings and precautions. This is particularly important when the recommended agent is a new and/or infrequently employed drug.

Worldwide distribution by S. Karger AG, Basel (Switzerland).
Printed in Switzerland by Médecine et Hygiène, Genève.
ISBN: 978-3-8055-8321-3

The scientific integrity of the articles contained in this volume is the sole responsibility of the authors.

First International Conference of the OIE Reference Laboratories and Collaborating Centres

•••••••••••••••••

Developments in Biologicals

Vol.128

This series"Developments in Biologicals" begins vith vol.102 and is the continuation of IABS symposia series "Progress in Immunobiological Standardization", Vol 1-5, "Immunobiological Standardization", Vols 1-22 and "Developments in Biological Standardization", Vols 23-101.

Organized and published by the International Association for Biologicals (IABS)

KARGER

Basel · Freiburg · Paris · London · New York
Bangalore · Bangkok · Singapore · Tokyo · Sydney

Steering Committee

A. Schudel (OIE Consultant)
L. Barcos (OIE)
G. Brückner (OIE)
J. Caetano Junior (MAPA, Brazil)
V. Caporale (IZS, Italy)
S. Edwards (VLA , UK)
B. Hill (CEFAS , UK)

Supporting Commitee

I. Bergmann (Panaftosa, Brazil)
F. Berthe (UPEI, Canada)
M. Bonnet (MAPA, Brazil)
I. Capua (IZS, Italy)
A. Camus (CIRAD-EMVT, France)
P. Dehaumont (AFSSA, France)
A. Diallo (FAO/IAEA, Austria)
M. Jeggo (AAHL, Australia)
A. Kida (HU, Japan)
D. Lightner (UA, USA)
T. Mettenleiter (FLI, Germany)
J. Mumford (Cambridge, UK)
T. Musoke (OVI, South Africa)
A. Nicola (SENASA, Argentina)
D. Paton (IAH, UK)
J. Pearson (USA)
B. Schmitt (NVSL, USA)
P. Vannier (AFSSA, France)
P. Wright (CFIA, Canada)
C. Zepeda (USDA-APHIS, USA)

Organisation of the Scientific Programme

Opening

Session I	OIE Reference Laboratories and Collaborating Centres: Mandates and Rules
Session II	Application of Laboratory Practices
Session III	Current Status and Approval of Laboratory Networking
Session IV	New Approaches
Session V	The Way Forward
Session VI	General Recommendations

Acknowledgements

The OIE is grateful for the invaluable support of the Ministry of Agriculture of the Federal Republic of Brazil, without which the First International Conference of the OIE Reference Laboratories and Collaborating Centres would not have been possible. The OIE also acknowledges the kind collaboration of the State of Santa Catarina, Brazil.

Opening

Introductory Speeches

An Historic Event

Dr Bernard Vallat

Director General of the World Organisation for Animal Health, Office International des Épizooties (OIE)

Honourable Governor,
Honourable Secretary,
Dear OIE Delegate of Brazil,
Dear OIE Experts and Delegates,
Honourable Participants,

It is a great pleasure for me on behalf of the President of the International Committee and all OIE Member Countries to welcome the First Conference of OIE Reference Laboratories and Collaborating Centres.

This is a truly historic event. It is the first time that the best scientists from the veterinary scientific community from all continents have come together. I would like to begin by giving my warmest thanks to the Brazilian Government and our Brazilian colleagues from the Federal and local Veterinary Services, the Brazilian Industry of Animal Production and our Regional Representation for the Americas for all the financial and technical contributions they have made to make this exceptional event possible.

As you know the OIE is the only intergovernmental organisation entirely dedicated to animal health worldwide. The objective of our Fourth Strategic Plan is: "to improve animal health worldwide".

We are committed to providing each and every one of our 168 Member Countries with the best methods of preventing and controlling animal diseases.

Our task is to collect, analyse and publish the updated scientific information in this field. We prepare, democratically adopt and publish standards and guidelines to ensure the safety of international trade in animals and animal products, and we provide guidelines for the surveillance of the more important animal diseases, including zoonoses.

The very high quality of input provided by the OIE to the veterinary community worldwide is linked to the activities of the network of our Reference Laboratories and Collaborating Centres.

You are at the heart of all OIE achievements. It is your constant input that makes it possible for the OIE to do its job. Your enormous potential to help our planet face existing and new threats can be greatly enhanced if together we now adopt common strategies.

As you know, our world has become more and more globalised. The movement of people and commodities is unprecedented in history, as well as the movement of pathogens and the cross-genetic opportunities that can arise in this context.

Climatic changes allow vectors to colonise new territories. If we are to tackle this new situation, new strategies need to be discussed and implemented not only at the national and regional levels, but mainly at the global level.

This is a great common responsibility that we all have to face together. This will not be possible if we do not start by improving the governance of Veterinary Services, by updating legislation and the quality and efficiency of all the potential tools used to implement legislation.

The quality of diagnostics and vaccines, as well as veterinary research, are key components of good governance and the official mandate of the OIE, through our Fourth Strategic Plan, is to influence Governments and international financial organisations to adopt new strategies and to give more resources so that these objectives can be achieved.

But in our globalised world, the failure of just one single country to implement early detection and rapid response when faced with an emerging or re-emerging disease outbreak now poses a threat to the entire planet. This is why solidarity has become a key word in our official objectives.

Investing more in the better governance of national and regional animal health systems will have a very strong impact on the alleviation of poverty, regional and international market access and, of course, public health, including food safety.

This investment will also protect rich countries that are free of diseases from importing pathogens. In the context of globalisation, border protection is no longer a sufficient method of protecting national territories. Investing in the support given to developing or transition countries not yet free of diseases is now the best cost/benefit approach to protecting disease-free countries.

It is not possible to help poor countries control or prevent animal diseases if they cannot use their own veterinary scientific community. The development of a scientific veterinary community in poor countries is also a priority, not only for the prevention and control of animal diseases at the national level, but also to allow those countries to participate more in the discussions and preparation of international standards to be adopted by the OIE; all this gives them a chance to participate more in world trade.

In the past, external support to the scientific community of poor countries through grants or loans provided to Governments, Ministries or national agencies by donors has often given very disappointing practical results, mainly because bureaucracy is not efficient and consumes too much money.

This is why the OIE developed the concept of twinning between existing OIE Reference Laboratories and Collaborating Centres and candidates from developing and transition countries wishing to become OIE Reference Laboratories.

This twinning concept will be presented and discussed during the Conference. It will allow direct cooperation between laboratories in a better cost-efficient way. It will contribute to the reinforcement of the OIE network and the veterinary scientific community worldwide.

Finally, I would like to announce that our strategy of networking and support to the OIE Reference Laboratories and Collaborating Centres will lead us to propose other meetings with you. This is the first but not the last one.

I would like to thank you again for coming. As with any event of such vision, translating good ideas into good outcomes often creates problems that must be overcome. If you have encountered such problems, let me personally thank you for your commitment, patience and understanding. It is our first meeting, we are learning, the next one and those that follow will be perfect.

My last message is again directed to our colleagues in Brazil. It is their unique and exceptional financial and technical contributions that have allowed us to be together in this paradise and I ask you to give them your wholehearted applause.

Discurso Acolhedor

Gelson Sorgato

Secretário da Agricultura e Desenvolvimento Rural de Santa Catarina

Quero cumprimentar, em nome do Governador Eduardo Pinho Moreira e da população do Estado de Santa Catarina, os participantes da 1ª Conferência Internacional de Laboratórios de Referência e Centros Colaboradores da Organização Mundial de Saúde Animal.

A escolha da nossa capital para sua realização, assim como da 18ª Conferência da Comissão Regional da OIE para as Américas, se constitui em motivo de orgulho e também de reconhecimento ao trabalho do povo desta terra.

O Estado de Santa Catarina, com uma população de seis milhões de pessoas e uma área geográfica de 95 mil quilômetros quadrados, no Sul deste imenso país, representa apenas 1,12% do território brasileiro, porém, faz parte de uma região de expressivo desenvolvimento social e econômico.

Sua população é formada por imigrantes europeus, especialmente portugueses, italianos, alemães, poloneses, franceses, austríacos, espanhóis e holandeses que aqui chegaram em meados do século 19 e início do século 20.

Esses imigrantes integrados aos brasileiros daquela época criaram uma maravilhosa cultura, que resultou num excelente desenvolvimento social e econômico.

A diversificação das atividades do setor agropecuário constitui a base e a garantia de um contínuo desenvolvimento das 200 mil pequenas propriedades rurais, já que 90% delas possuem até 50 hectares.

Na produção animal, o Estado de Santa Catarina é responsável por 13% das carnes brasileiras. Está produzindo neste ano 700 mil toneladas de carne suína e 1 milhão e meio de toneladas de carne de frango, além de 125 mil toneladas de carne bovina e 1 milhão e meio de toneladas de leite.

Santa Catarina possui o maior e mais moderno parque industrial de frigoríficos de suínos e de aves do país. Participa de importantes mercados internacionais na condição de fornecedor de alimentos de boa qualidade e com segurança alimentar.

As exportações catarinenses de carne suína representam 45% das exportações brasileiras, enquanto 20% da carne de frango exportada pelo país têm origem neste estado.

Atualmente se observa um progressivo aumento da preocupação da população mundial com a segurança sanitária dos alimentos, representando inclusive alterações no modelo de consumo alimentar a exemplo dos reflexos da ocorrência da Influenza Aviária e da Encefalopatia Espongiforme Bovina.

Neste sentido, a Organização Mundial de Saúde Animal possui uma grande atribuição e extrema responsabilidade. Sabe-se que é esta organização que promove a segurança sanitária dos animais e de seus produtos para alimentação humana em escala mundial. Sabe-se que a OIE possui a melhor equipe técnica do mundo para promover o diagnóstico laboratorial, o controle e a erradicação das enfermidades dos animais, com reflexos diretos na saúde humana.

Cada vez mais aumenta a importância das atividades da OIE na segurança alimentar de um mundo cada vez mais globalizado, considerando não somente os agentes patogênicos e o bem-estar dos animais, mas também o uso indiscriminado de produtos pesticidas e outros produtos de síntese química.

Pela primeira vez a OIE está reunindo sua rede laboratorial formada por 160 laboratórios e 20 Centros Colaboradores do mais alto nível científico, através da participação das senhoras e dos senhores, e para nossa alegria, escolheu para esta 1ª Conferência o Brasil e em especial, o Estado de Santa Catarina e sua capital Florianópolis.

As informações sobre esta rede laboratorial e Centros Colaboradores, em matéria de assessoramento técnico e científico sobre temas de vigilância sanitária e de luta contra enfermidades no mundo, sinalizam para uma segurança sanitária extraordinária para a criação de animais, para com isto garantir a oferta de alimentos de origem animal de boa qualidade e com total garantia sanitária para a população humana.

Os catarinenses reconhecem a importância e valorizam o prestígio da Organização Mundial de Saúde Animal, razão pela qual o Estado de Santa Catarina vem se preparando durante muitos anos para receber desta entidade o reconhecimento de Zona Livre de Febre Aftosa Sem Vacinação, único estado brasileiro com esta condição sanitária.

Vacinamos o rebanho bovino durante 33 anos. O último foco de febre aftosa em terras catarinenses ocorreu há 14 anos e, há seis anos e meio, estamos mantendo o rebanho de 3 milhões de cabeças sem vacinação.

O estado está concluindo os procedimentos junto ao Ministério da Agricultura do Brasil, para apresentação do pedido desta certificação no período que antecede a Reunião Anual da OIE em Paris, no próximo mês de maio.

Portanto, é grande a expectativa dos criadores, das agroindústrias, do governo estadual para que o Estado de Santa Catarina receba, em 2007, a certificação desta entidade mundial, contribuindo para uma maior participação no mercado internacional, especialmente de carne suína e de aves.

Para finalizar, quero destacar a admiração e o respeito ao Dr. Bernard Vallat, Diretor Geral da OIE e ao Dr. Barry O'Neil, presidente do Comitê Internacional da OIE, que conduzem esta entidade, com a participação de todos os delegados dos países membros e com a cooperação científica da rede de Laboratórios e Centros Colaboradores.

Quero destacar também a grande importância do Ministério da Agricultura, Pecuária e Abastecimento do nosso país. O ministro Luís Carlos Guedes Pinto e o Dr. Jamil Gomes de Souza, vem permanentemente prestigiando e contribuindo para o desenvolvimento da agropecuária catarinense, assim como de todo o país.

Em nome do Governador do Estado e de todos os catarinenses, desejo para as senhoras e para os senhores participantes da 1ª Conferência Internacional de Laboratórios de Referência e Centros Colaboradores da OIE, uma excelente permanência no Estado de Santa Catarina, assim como desejo sucesso e votos de grandes realizações em benefício da população mundial.

Opening Address

Dr Barry O'Neil

President of the World Organisation for Animal Health – OIE

Distinguished guests, friends and colleagues, ladies and gentlemen,

It is extremely pleasing for me as President of OIE to be present for the first conference of OIE's Reference Laboratories and Collaborating Centres. And I would like to start my short presentation by giving a special thanks to the government of Brazil for making this conference possible.

To me the 160 OIE Reference Laboratories and the 20 Collaborating Centres, for more than 80 diseases of terrestrial and aquatic animals, are the foundation that enables OIE to be successful as the world organisation for animal health.

The OIE is a science-based organisation setting standards that are based on the best science available. And most of the science used comes from the more than 250 internationally renowned experts, located in over 30 countries that host the reference laboratories and collaborating centres.

And I would like to express my thanks to the host countries of these laboratories and centres, for making available their facilities and experts to help the OIE in the way they do.

The OIE is very fortunate to have such expert support, so it is pleasing for me that at last, with the help of the government of Brazil, we have been able to bring you all together here, at the wonderful location of Florianopolis.

Over the course of the next 2 days, I hope you will be able to significantly improve your networks and links to each other, and also be updated on relevant new developments. Conferences like this succeed as a result of the engagement and commitment of all those involved, so I hope you will make the most of the opportunities presented by this event.

There are many challenges, I am sure, to overcome in your respective areas, but I honestly believe that working collaboratively will enable us to be best placed to face and address these challenges.

This afternoon I briefly sat in on the aquatic workshop with Barry Hill and other aquatic experts. Listening to the discussions made me very aware of some of the challenges in the aquatic area, including those created by wonderful new technology. This means that additional strains of agents and pathogens can now be identified - something that was not previously possible. But this has presented a challenge, in that some of these agents may not be disease causing (either currently or in specific situations), resulting in uncertainty about what if anything should be done to stop these agents spreading.

However without good science input into these types of issues, policy makers usually make poor decisions. Even in the face of scientific uncertainty, decisions still need to be taken, and these need to be made based on whatever sound science that is available, as well as ensuring that where there are significant information gaps, the work is underway to fill these gaps.

Turning now to the Collaborating Centres, I would like to acknowledge the role that these important groups play in training and advising countries and experts on specific issues. In this area the needs of the world are huge, especially in developing countries, and the support we can collectively give these countries is critical for them to improve their situations.

In my opinion we all have a moral obligation to do what we can to help these countries, within our respective areas of expertise.

Finally it is also pleasing to me that the OIE has specifically recognised in its Fourth Strategic Plan the critical role played by Reference Laboratories and Collaborating Centres. In developing this strategy we looked at ways that we can support you more in doing the important work that you do. I hope this 1st International Conference is a positive first step, and I look forward to discussing with you over the next two days, what the OIE can further do to support your future activities.

Thank you all for making the effort to participate in this conference, and I wish you every success for not only the activities over the next two days, but also for your work with the OIE in the future.

Welcoming Address

Dr Steven Edwards

President of the OIE Biological Standards Commission

I would like to thank all the Delegates for coming to Florianopolis.

This is the first ever such event, namely a conference that is focused on the role and activities of OIE Reference Laboratories and Collaborating Centres. There are many scientific conferences worldwide, particularly relating to research studies, which may be more or less specialised to particular diseases or to particular disciplines. There are also conferences dealing with disease control and epidemics, and even a few with a primary focus on diagnostics. This meeting however, is unique. I have worked with colleagues at the OIE over the past 15 years as the concept of OIE Reference Laboratories was first promulgated, and subsequently developed and refined. Then the concept of Collaborating Centres also emerged, not so much aimed at specific diseases and their diagnosis and control, but rather taking cross-cutting topics and providing centres of knowledge, expertise and training for the wider community.

So welcome to you all. Whilst we have many distinguished speakers, it is my hope that your lasting memories of the conference will be the opportunities it provides to talk to each other informally, to build personal relationships and to take every opportunity to extend your networks.

This conference could not have happened without the vision of the OIE, particularly the Director General, Dr Bernard Vallat, and the recently retired Head of the Scientific and Technical Department, Dr Alejandro Schudel.

Above all we are all extremely grateful for the most generous sponsorship of the Government of Brazil without which we would not be here today.

Contents

Lombard M, Dodet B, the OIE Scientific & Technical Department (eds): First International Conference of the OIE Reference Laboratories and Collaborating Centres. Dev Biol (Basel). Basel, Karger, 2007, vol 128, pp 1-2.

Introduction

Bernard Vallat

Director General, OIE

The World Organisation for Animal Health (OIE), with the valuable support of the Ministry of Agriculture of the Federal Republic of Brazil, organised the "First International Conference of the OIE Reference Laboratories and Collaborating Centres", which was held in Florianopolis, Brazil, from 3 to 5 December 2006. Representatives from all the OIE Reference Laboratories and Collaborating Centres, and Chief Veterinary officers as well as national scientific experts and private laboratory experts, were invited to participate in the Conference.

The purpose of the Conference was to provide a multidisciplinary forum for strengthening the scientific cooperation within the network of OIE and FAO Reference Laboratories and OIE Collaborating Centres. The conference further promoted the updating and setting of standards for methodologies in the field of diagnostics, vaccine quality and biosecurity; the improvement of links between existing Reference Laboratories, Collaborating Centres and national official and private laboratories; and support to developing and in-transition countries through capacity building and training programmes and the identification of initiatives for financial support.

Over the past few years the world has witnessed a global emergence and re-emergence of several infectious animal diseases that have had, and are still having, significant impact on both animal and human health. These infections have not only caused much concern to governments, policy makers and scientists all over the world, but some of them have posed significant threats to the livelihoods of the farming community, especially in rural areas. Among several of the limitations to stopping the spread of these diseases is the lack of appropriate reference laboratories in many developing countries for the early detection of, reaction to and timely notification and control of disease outbreaks. This issue has been particularly addressed in the design of the Fourth Strategic Plan of the OIE, taking into consideration the expertise and capabilities developed over time by OIE

Member Countries in establishing a unique worldwide network of OIE Reference Laboratories and Collaborating Centres.

Through its four Specialist Commissions, 160 Reference Laboratories and 20 Collaborating Centres for more than 80 diseases of terrestrial and aquatic animals, located in 30 countries, and in collaboration with more than 250 internationally renowned experts, the OIE conducts its actions and decisions based on scientific justification as required in the Agreement on the Application of Sanitary and Phytosanitary Measures (SPS Agreement) of the World Trade Organisation (WTO). The OIE is recognised within the SPS Agreement as the only International Organisation responsible for setting international standards, recommendations and guidelines for the trade in animals and animal products.

The policy of the OIE and its partner organisation FAO, is to seek methods for international, regional and national support, to strengthen and improve the delivery of veterinary services worldwide including the strengthening of activities and capacities of veterinary laboratories.

The OIE acknowledges with appreciation the support of the Government of Brazil, for collaborating on the implementation of this initiative, and hopes that the outcomes of this First Conference will benefit the world animal health community in serving towards better control and prevention of animal diseases and zoonoses.

Lombard M, Dodet B, the OIE Scientific & Technical Department (eds): First International Conference of the OIE Reference Laboratories and Collaborating Centres. Dev Biol (Basel). Basel, Karger, 2007, vol 128, p 3.

Session I

OIE Reference Laboratories and Collaborating Centres: Mandates and Rules

Lombard M, Dodet B, the OIE Scientific & Technical Department (eds): First International Conference of the OIE Reference Laboratories and Collaborating Centres. Dev Biol (Basel). Basel, Karger, 2007, vol 128, pp 5-11.

The Work of the OIE Biological Standards Commission

S. Edwards

President of the Biological Standards Commission, Veterinary Laboratories Agency (Weybridge), UK

Key words: Infectious diseases, international standardisation, diagnostic tests, vaccine standards, OIE Terrestrial Manual

Summary: Founded in 1949 and now composed of five specialist members, this Commission establishes or approves methods for diagnosing diseases of mammals, birds and bees and for testing biological products, such as vaccines, used for control purposes. It oversees the production of the *Manual of Diagnostic Tests and Vaccines for Terrestrial Animals* (the *"Terrestrial Manual"*), recognised as an international standard text under the SPS Agreement of the World Trade Organisation. The Commission also selects OIE Reference Laboratories for diseases of terrestrial animals, and promotes the preparation and distribution of standard reagents for diagnostic testing. Great reliance is placed on the OIE network of expertise, and specialists are brought together in Ad hoc groups to address specific issues in relation to standardisation of diagnostic tests and vaccines.
The Commission is elected by the International Committee for a three year term. Commission Members should be recognised specialists, with international experience, trained in infectious disease diagnosis and/or prevention, particularly in the area of laboratory methods.

BACKGROUND

The Office International des Epizooties (OIE) established a "Norms [Standards] Commission" in 1949 to develop internationally agreed standards for laboratory diagnostic tests and vaccines for OIE-listed animal diseases. Members of the Commission are elected for a 3 year term by the International Committee of the OIE. It was renamed the Biological Standards Commission (BSC) in 2003, and in 2006 the number of elected members increased from three to five. This reflects the steadily growing responsibilities of the Commission and the need to include a broad range of scientific expertise and geographical awareness within the group. An important element in the Commission's work is its ability to call upon other experts to assist at its meetings and to make full use of the expertise available in the OIE Collaborating Centres and Reference Laboratories. The full terms of reference are available on the OIE website (http://www.oie.int/bsc/eng/en_bsc.htm) but in summary its main tasks are to:

- establish or approve methods for diagnosing diseases of mammals, birds and bees and for manufacture and testing of biological products, such as vaccines; and advise OIE on appropriate use of diagnostic tests and vaccines;
- oversee production of the *Manual of Diagnostic Tests and Vaccines for Terrestrial Animals*;
- select OIE Reference Laboratories for diseases of mammals, birds and bees; communicate with these laboratories as a source of specialist expertise;
- promote the international standardisation of diagnostic tests, including the preparation and distribution of standard reagents.

The Commission meets formally in January and September each year, but business continues between meetings through email exchanges and correspondence. The Commission can ask the Director General of the OIE to set up Ad hoc groups to address specific technical and scientific issues, and such groups report back to the BSC. The President of the Commission reports to the International Committee of the OIE each year in May, including reports from the Ad hoc groups. At that time any recommendations of the Commission or new standard texts are put forward for formal adoption.

The OIE Terrestrial Manual

The importance of the Commission's work gained a new emphasis in 1989 with the publication of Volume I of the *OIE Manual of Recommended Diagnostic Techniques and Requirements for Biological Products.* This comprised individual disease chapters drafted by scientific experts, collated by the Standards Commission and endorsed by the International Committee of the OIE. Volumes II and III (all loose leaf) followed in subsequent years. The 2nd edition, published in 1992 under the title of *Manual of Standards for Diagnostic Tests and Vaccines*, was bound in a single volume, with subsequent editions every four years. The current (5th) edition, entitled *Manual of Diagnostic Tests and Vaccines for Terrestrial Animals* [1] was published in 2004 in two hard bound volumes.

The recognition of OIE by the World Trade Organisation as the international standards-setting body for animal health has brought the *Terrestrial Manual* into new prominence. The OIE's *Terrestrial Animal Health Code* [2] describes international trade and surveillance standards for terrestrial animals and their products, and refers to the *Terrestrial Manual* for standards and procedures for diagnostic tests and vaccines. 'Prescribed tests' defined in the *Terrestrial Manual* are considered appropriate to test animals that are moving internationally. Therefore, the *Code* and the *Manual* are companion standards. In addition to prescribed tests, the

Manual also describes 'alternative' tests. These assays are suitable for the detection of diseased or infected animals within a local setting, and can be used in the import/export of animals but only after bilateral agreement.

The scope of the *Terrestrial Manual* includes a series of generic introductory chapters of relevance to veterinary laboratories, with a chapter on each of the OIE-listed diseases plus a number of other important diseases. Each disease-specific chapter starts with a general introduction to the disease, then a description of procedures for laboratory-based diagnosis, and finally, where relevant, a section on standards for vaccine production.

Preparation of the *Terrestrial Manual* is a major task for the BSC. An outline of the process is given below:

- Individual chapters are commissioned from international experts on the dis ease or topic. Where possible these experts are selected from the OIE Reference Laboratories or Collaborating Centres.
- Draft texts are sent to all Member Countries, peer reviewers, and relevant Reference Laboratories
- The consultant and scientific editors prepare final texts incorporating comments received. Any issues of likely controversy, or where Member Country and expert comments are not concordant, are discussed with the BSC for a proposed resolution.
- The texts are presented for adoption by the OIE International Committee.

Because some chapters need updating between the 4-yearly printed editions, a mechanism has been agreed whereby individual chapters adopted by the Committee in intervening years will be placed on the OIE Website. Thus the most up to date version of the *Terrestrial Manual* is always the web edition.

Reference Laboratories and Collaborating Centres

The BSC has developed guidelines for candidate institutions wishing to become designated as an OIE Reference Laboratory or Collaborating Centre. These require the institution to demonstrate a track record of scientific activities relevant to the selected disease or topic, including evidence of existing international activities. The mandate of Reference Laboratories requires them to function as a centre of expertise and standardisation, to prepare, store and distribute biological reference materials, to develop new procedures for disease diagnosis and control, to gather, analyse and disseminate epidemiological data, and to provide expert consultancy for the OIE. A number of ancillary activities are also desirable such as training, diagnostic testing for other countries, organisation of scientific meetings, and publication of scientific data. Collaborating Centres are focused on cross-cutting topics rather than specific diseases, and they are not necessarily laboratory-based (although many are). Their mandate relates to operating as a centre of research, expertise, standardisation and dissemination of techniques, the harmonisation of international regulations for animal disease surveillance and control, and provision of consultant expertise. Again a number of ancillary activities are also desirable such as training, scientific meetings and publications.

Apart from aquatic animal diseases (which are dealt with by the Aquatic Animals Commission), the BSC assesses applications for Reference Laboratory and Collaborating Centre status, and advises the OIE on their suitability. For established Reference Laboratories and Collaborating Centres, the BSC requests an annual report of their activities. These reports are collated, analysed and published by the OIE each year. The status is reviewed every four years and may be withdrawn if there is insufficient evidence of relevant international activity. The OIE is keen to establish more Reference Laboratories in developing countries, but this must not be at the expense of scientific competence. To encourage this process a twinning scheme is being proposed to foster capacity building in developing country laboratories with direct inputs from established laboratories in developed countries. Where there is more than one OIE Reference Laboratory for a given disease, the BSC actively encourages them to interact and, for high profile topics such as foot and mouth disease and avian influenza, formal networks have been established.

Guidelines and Quality Standard

Some years ago, the BSC recognised there was a need for some practical guidelines to assist laboratories involved in test standardisation, and also to enable national reference laboratories and international OIE Reference Laboratories to achieve a recognised standard of quality accreditation. These were eventually compiled and published in a booklet [3] covering guidelines for test validation, preparation of biological reference materials for antibody assays, laboratory proficiency testing, and the OIE *Standard for Management and Technical Requirements for Laboratories Conducting Tests for Infectious Animal Diseases.* The last is a specific interpretation of the more generally stated requirements of the ISO/IEC 17025:1999 international quality standard for testing laboratories. The OIE quality standard clearly states that test methods and related procedures must be appropriate for specific diagnostic applications in order for the test results to be of any relevance. In other words, the assay must be ‘fit for purpose’. The Quality Standard further states that in order for a test method to be considered appropriate, it must be properly validated and that this validation must respect the principles outlined in the validation chapters of the Manual. The guidelines and quality standard are currently being revised and updated.

Standardisation of Diagnostic Tests and Vaccines

International standardisation of laboratory procedures is a huge task, and a very important one in the era of globalisation. The BSC aims to focus first on the higher profile topics, as well as encouraging individual Reference Laboratories to initiate their own projects as part of their mandate. The key concept is for all the key laboratories to agree on a "standard reference method" to which all others can be compared. This in itself is challenging because different scientists will favour different methods. A good starting point is therefore to set up an international comparison of different methods on a standard panel of test samples. Eventually a reference method can be agreed that can be published in the OIE *Manual*. Linked to this is the establishment of internationally agreed "standard reference reagents" – typically sera, using the OIE Guidelines to prepare them. The OIE has already recognised reference reagents for 15 diseases, and these are listed on the OIE website (http://www.oie.int/eng/normes/en_sera.htm) together with the name of the supplier. Work on other diseases is ongoing. One of the problems faced by Reference Laboratories is obtaining funding for such programmes, as it tends to fall between recognised activities such as research or surveillance.

In the field of vaccines much less progress has been made, mainly on equine influenza where the OIE Reference Laboratories have prepared and characterised a set of antisera for vaccine antigen calibration. They have also established an expert panel that meets annually to advise OIE and vaccine manufacturers on the currently circulating strains of equine influenza virus that should be included in commercial vaccines.

Validation of Diagnostic Tests and the OIE Register

As mentioned above, the move towards formal quality accreditation in diagnostic laboratories has driven a requirement to demonstrate that tests in use are validated and fit for purpose. Beginning with the Guidelines booklet [3] and then through successive revisions of the validation chapters in the *OIE Manual* [1], a well recognised system has been developed for validating veterinary diagnostic tests for infectious diseases.

While the test methods currently described in the *Manual* are considered to be standard methods, many of them are 'classic' in that they have been used in diagnostic laboratories for many decades and their performance characteristics, including limitations, are well known. These assays, while still useful, will remain *status quo* in terms of reagents and format until replaced with newer technologies. In future, new methods for inclusion in the *Manual* will be required to have met comprehensive validation criteria, including their suitability for use as 'prescribed tests' for international trade, or for other purposes such as disease diagnosis of clinical cases or disease surveillance.

Working with the OIE Collaborating Centre for ELISA and Molecular Techniques in Animal Disease Diagnosis, based at the FAO/IAEA Joint Division in Vienna, Austria, together with representatives of the diagnostics industry, the OIE has recently established a Register of diagnostic tests that will be published on the

OIE website (http://www.oie.int/vcda/eng/en_background_VCDA.htm). This is aimed specifically at diagnostic kit manufacturers who by this means can demonstrate that a test has been comprehensively validated, assessed by a panel of international experts, and designated fit for the specified purpose. A more detailed account of the establishment of the Register has been published by Wright et al. [4].

Expert Ad Hoc Groups

Even within the world of infectious animal disease diagnostics and vaccines, the complexity of the science and the rapid advances in technology mean that one small committee cannot possibly encompass the expertise needed to fulfil the expectations of the OIE and its delegates. The BSC therefore places great reliance on Ad hoc groups, convened by the Director General of the OIE for a specific purpose. The reports of such groups are considered by the relevant specialist commission (such as the BSC) and, if acceptable, published and presented to the OIE International Committee along with the report of the Commission. Some recent examples relevant to the work of the BSC include:

- non-structural protein tests for foot and mouth disease;
- tests for bovine spongiform encephalopathy;
- biosafety and biocontainment standards for laboratories;
- antimicrobial resistance;
- biotechnology.

Concluding Remarks

As well as the major activities outlined above, members of the BSC regularly represent OIE at international conferences, provide technical advice through correspondence, and ensure that a good liaison is maintained with the other OIE Specialist Commissions. Inevitably, because much of the work centres around laboratory science, and is poorly understood in the wider world of animal health and disease control, the BSC has also an important role in communication, which needs to be further developed in the future. A key success factor for the BSC has been the establishment of the network of Reference Laboratories and Collaborating Centres. The development of this network and its geographical extension will be an important aim for the future.

Acknowledgements

I am very grateful to all the members of the Biological Standards Commission, past and present, for their valued and expert support. Other regular contributors to our discussions include Peter Wright and Adama Diallo. Jim Pearson is the consultant editor to the *Terrestrial Manual*, without whose efforts it would never see the light of day. All the staff of the Scientific and Technical Dept at the OIE provide constant and high quality support, but I would particularly single out Sara Linnane for her excellent editorial skills and good humour.

References

1. Office International des Epizooties (OIE): Manual of Diagnostic Tests and Vaccines for Terrestrial Animals, 5th Ed. OIE, Paris, 2004, 1178 pp.
2. Office International des Epizooties (OIE): Terrestrial Animal Health Code, 15th Ed. OIE, Paris, 2006, 652 pp.
3. Office International des Epizooties (OIE): OIE Quality Standard and Guidelines for Veterinary Laboratories, OIE, Paris, 2002, 63 pp.
4. Wright P, Edwards S, Diallo A, Jacobson R: Development of a framework for international certification by OIE of diagnostic tests validated as fit for purpose. Dev Biol (Basel) 2006, 126, 43-51.

Pr. Steven Edwards, Veterinary Laboratories Agency, New Haw, Addlestone, KT15 3NB, UK
E-mail: s.edwards@vla.defra.gsi.gov.uk

Lombard M, Dodet B, the OIE Scientific & Technical Department (eds): First International Conference of the OIE Reference Laboratories and Collaborating Centres. Dev Biol (Basel). Basel, Karger, 2007, vol 128, pp 13-20.

The Work of the OIE Aquatic Animal Health Standards Commission

B. Hill

Vice-President of the Aquatic Animal Health Standards Commission, Centre for Environment, Fisheries and Aquaculture Science, Weymouth, UK

Key words: Aquatic animals, infectious diseases, international standards, Aquatic Code, Aquatic Manual.

Summary: Founded in 1960 as the Fish Diseases Commission, the Aquatic Animal Health Standards Commission is currently composed of five members elected by the OIE International Committee on a three-year basis. Its remit covers diseases of fish, molluscs and crustaceans. OIE stipulates that Commission members should be internationally recognised specialists in the fields of methods for surveillance, diagnosis and prevention of infectious aquatic animal diseases and have extensive international experience, at the regional or global level. The Commission is responsible for developing the international standards of the *Aquatic Code* and the *Aquatic Manual*, very ably supported by various OIE Ad hoc Groups and the designated experts at the OIE Reference Laboratories for aquatic animal diseases. The latest editions of the *Aquatic Code* and the *Aquatic Manual* published in 2006 incorporate several important modifications including several changes to the list of diseases. Work has commenced in new areas such as aquatic animal welfare and an assessment of whether amphibian diseases should be included in the work of the Commission. Continuing efforts are being made to encourage enhanced involvement of veterinary authorities in aquatic animal health and to improve cooperation between veterinary and other authorities with competence for aquatic animal health.

INTRODUCTION

Aquatic animal health is of increasing importance because of the substantial worldwide growth taking place in the volumes and values of fish, molluscs and crustacean species, farmed both for essential food supply at the local level and for export trade at the national level. Disease outbreaks are causing significant losses in aquaculture production throughout the world and are having a detrimental effect on valuable export trade in aquaculture animals and their products, particularly for some developing countries. Although local pathogens, inadequate farm-management, environmental factors and poor water quality are the most common causes of disease outbreaks in farmed fish and shellfish, pathogen transfer due to international trade in live aquatic animals or their products continues to be a major reason for new disease occurrences in importing countries. For this reason, the OIE standards cover diseases and international trade of aquatic animals as well as terrestrial animals.

In 1960, the Office International des Epizooties (OIE) established the Fish Diseases Commission because of the increasing awareness of the importance of the spread of infectious diseases of fish through increasing international trade in fish. In 1988, the scope of the Commission was extended to include diseases and pathogens of molluscs and crustaceans, and in 2003 the Commission was renamed as the Aquatic Animal Health Standards Commission (in short, Aquatic Animals Commission). It is one of the four Specialist Commissions of the OIE. The role of Specialist Commissions is to study specific problems relating to the epidemiology and control of certain diseases or groups of diseases, and to report on their activities to the OIE International Committee, to which they also submit recommendations.

The five members of the Aquatic Animals Commission are elected for a three year term by the International Committee of the OIE at the General Session in May of the election year. The OIE requires that the Commission members be internationally recognised specialists in the fields of methods for surveillance, diagnosis and prevention of infectious aquatic animal diseases and have extensive international experience at the regional or global level. The full terms of reference and the mandate of the Commission, and details of its current members, are available on the Aquatic Animals Commission's pages on the OIE website (www.oie.int/aac/eng/en_fdc.htm). Key functions of the Commission are to:

- propose the most appropriate methods for surveillance, diagnosis and disease prevention for sanitary security of trade or international movement of aquatic animals and their products covering diseases listed in the *Aquatic Animal Health Code* (the *Aquatic Code*);
- oversee the updating and other revisions of the *Aquatic Code* and A*quatic Manual;*
- promote dissemination amongst Veterinary and other Competent Authorities of information on aquatic animal diseases;
- keep the International Committee and the Director General informed of scientific progress on methods for surveillance, diagnosis and disease prevention likely to improve the prevention and the control of aquatic animal diseases;
- identify issues that require in-depth review and propose, to the Director General, the composition and terms of reference of experts or Ad hoc Groups of experts convened specifically to study such issues, and if necessary, to participate in the work of these Groups;
- assess applications for appointment as OIE Reference Laboratory and designated expert, and advise the International Committee on their suitability;
- facilitate and work with the worldwide network of OIE reference laboratories

and collaborating centres in the field of aquatic animals, so as to achieve OIE's mandate.

The Commission meets formally at the OIE, Paris, in March and September/October each year but does much work by electronic communication between meetings. Internationally leading experts in aquatic animal health are asked to assist the Commission by attending its meetings as invited observers to provide additional specialist knowledge. In addition, the Commission may ask the Director General of the OIE to set up Ad hoc Groups to address specific technical and scientific issues, and such groups report back to the Commission with recommendations. The President of the Commission reports on its activities and new proposals to the International Committee at the OIE General Session in May, and includes any reports from the Ad hoc Groups. Any recommendations of the Commission for changes to the OIE standards are put forward for formal adoption. The adopted texts are made publicly available as soon as possible on the OIE web site.

Development of the OIE Standards for Aquatic Animals

The OIE standards for aquatic animals are prepared by the Aquatic Animals Commission with the assistance of internationally renowned independent experts, OIE Ad hoc Groups and expertise at the OIE Reference Laboratories for aquatic animal diseases. The views of OIE Member Countries are systematically sought through the circulation of draft and revised texts and other proposals appended to the Commission's meeting reports to the National Delegates. In addition, the Aquatic Animals Commission collaborates closely with the OIE Terrestrial Animal Health Standards Commission on issues needing a harmonised approach in the *Terrestrial Code* and the *Aquatic Code*, and with the Biological Standards and Scientific Commissions to ensure consistency of scientific standards in the two Diagnostic Manuals.

The main standards produced by the OIE for aquatic animals are the *Aquatic Animal Health Code* (*Aquatic Code*) and the *Manual of Diagnostic Tests for Aquatic Animals* (*Aquatic Manual*). In addition to being published as hard copy, the *Aquatic Code* and the *Aquatic Manual* are also made available electronically on the OIE web site under 'Health standards'. Because of scientific advances being made in this fast-moving field, some chapters in the *Aquatic Manual* require updating between the four-yearly printed edition so individual chapters adopted by the International Committee in intervening years are placed on the OIE Website. The most up to date version of the *Aquatic Manual* is the web edition.

The latest printed edition (9th) of the *Aquatic Code* [1] and the latest edition (5th) of the *Aquatic Manual* [2] incorporate modifications proposed by the Aquatic Animals Commission and approved by Delegates during the General Session in May, 2006.

Harmonising the Aquatic and Terrestrial Codes

The need to harmonise OIE approaches to aquatic and terrestrial animal diseases continues to be addressed by the Aquatic Animals Commission. While progress has been on-going in the areas of disease listing criteria, notification and epidemiological information and data entry forms, work still needs to be done, especially on zoning and compartmentalisation, aquatic animal health surveillance, model aquatic animal health certificates, and on disposal of carcasses and wastes of aquatic animals. This work will be carried out over the next two years.

Examples of other activities of the Aquatic Animals Commission

There are many issues addressed by the Commission other than developing the international standards for the OIE listed diseases. Some recent examples are as follows:

OIE standards for aquatic animal welfare

The International Committee has decided that the OIE should give priority to the welfare of animals used in agriculture and aquaculture, and that the topics of transportation, humane slaughter and killing for disease control purposes should be addressed first, followed by housing and management. Other topics, such as research animals and wildlife, will be addressed as resources permit.

A set of principles for the welfare of aquatic animals as well as four sets of proposed guidelines (slaughter of farmed fish for human consumption; humane killing of fish for disease control purposes; transport by land; and transport by sea) have been developed by two aquatic Ad hoc Groups and operated under the OIE Working Group on Animal Welfare. As a result of recommendations made by this Working Group, the principles for aquatic animal welfare were harmonised to the extent possible with the corresponding text contained in the Terrestrial Code. The proposed welfare chapters for the *Aquatic Code* were sent to Member Countries for their opinion. Detailed comments were received and considered by the Aquatic Animals Commission at its meeting in October 2006 and then submitted with the Commission's comments to the Ad hoc Groups and the Working Group on Animal Welfare for further refining of the chapters before circulating to Member Countries again for comment.

Improving co-operation between veterinary and fishery authorities for aquatic animal health

At the 72nd General Session in May, 2004, the International Committee endorsed a set of recommendations to improve cooperation between veterinary and fisheries authorities on aquatic animal health matters in order to facilitate the implementation of the international obligations of Member Countries.

Over the past two years, more than 40 OIE Member Countries have responded to the request from the Director General of the OIE to nominate a specific official to be the OIE contact person for aquatic animal diseases with responsibility for sending, under the authority of the Delegate, comments on the Commission's proposals for new standards, and reports of aquatic animal disease occurrence in their country. It is hoped that many more Member Countries will follow this lead, which is expected to enhance the awareness of more veterinary authorities to developments in aquatic animal health and lead to improved levels of disease reporting to the OIE.

Assessing whether amphibian diseases should be included in the remit of the OIE

The Aquatic Animals Commission has been requested by some Member Countries to include amphibian diseases in its scope and will be advised by the newly established OIE Ad hoc Group on Amphibian Diseases in assessing the case for doing so. The international trade in live amphibians and their products is a multi-million dollar industry, including trade of amphibians for human consumption, laboratory animals, pet-trade and zoo animals.

At its first meeting in September 2006, the Ad hoc Group amended the Questionnaire on amphibian diseases' distributed to Member Countries in 2002 but to which there was a very poor response. After endorsement by the Aquatic Animals Commission, the amended questionnaire was sent by the Director General to all Member Countries for response by February 2007. It is hoped that Member Countries will provide the information required by replying to the questionnaire so that the Aquatic Animals Commission can reach a decision on whether to propose that OIE standards should be developed for international trade in amphibian animals.

Listing and notification of diseases by strain/genotype of the pathogen

The Aquatic Animals Commission has prepared a position paper on possible listing and notification of diseases by strain/genotype, with examples of finfish, mollusc and crustacean diseases where differences in virulence have been documented for different strains/genotypes of the pathogenic agents of listed diseases. This formed the basis of a special workshop held just before this First International Conference of OIE Reference Laboratories and Collaborating Centres. The workshop participants reached a number of conclusions and produced the following three recommendations.

1. The issue of whether or not distinct genotypes of OIE listed diseases should be listed as such and have individual reporting requirements should be discussed in a wider forum at the next meeting of OIE Reference Laboratories.
2. Because different laboratories are using different methods for genotyping, OIE Reference Labs should be requested by the OIE to take the lead in developing and harmonizing methods for differentiation of genotypes.

3. The implications of differentiating between genotypes for OIE notification and reporting criteria should be considered by the Aquatic Animals Health Standards Commission.

OIE publication on aquatic animal disease emergencies

The Aquatic Animals Commission with the assistance of Australian experts is helping OIE to coordinate an issue of the OIE Scientific and Technical Review series entitled "Changing Trends in Managing Aquatic Animal Disease Emergencies" with publication planned for April 2008.

Expert Ad Hoc Groups

The field of aquatic animal health is of increasing complexity and importance. Although the members of the Aquatic Animal Commission have wide knowledge and expertise in the subject, they cannot cover all aspects, and so require the support of other experts working as Ad hoc Groups. These groups are convened as required by the Director General of the OIE to examine specialist scientific and technical issues. Comprising leading specialists from OIE Member Countries, their reports serve as guidance and advice to the Commission and the International Committee in forming recommendations and making decisions. Recent examples of areas covered by OIE Ad hoc Groups supporting the work of the Aquatic Animals Commission include:

- review of the OIE list of aquatic animal diseases;
- development of new chapters on fish, mollusc and crustacean diseases for the *Aquatic Code* and *Aquatic Manual;*
- surveillance guidelines chapters for the *Aquatic Code* and *Aquatic Manual*;
- standards for the slaughtering/killing of fish;
- standards for aquatic animal transport;
- assessment of whether amphibian diseases should come into the remit of the OIE; and
- standards for aquatic animal feed.

The contribution of the individual members of the Ad hoc Groups who selflessly give their time freely, despite their many other obligations, to provide expertise through electronic communication and attendance at Ad hoc group meetings is highly appreciated by the Commission and the OIE.

Reference Laboratories and Collaborating Centres

The role of an OIE Reference Laboratory is to function as a centre of expertise and help develop the standardisation of diagnostic techniques for its designated disease. The Designated Expert, responsible to the OIE and its Member Countries with regard to these issues, should be a leading and active expert helping the Reference Laboratory to provide scientific and technical assistance and specialist advice on topics linked to diagnosis, surveillance and control of the disease for which the Reference Laboratory is responsible. They may also provide scientific and technical training for personnel from Member Countries, and coordinate scientific and technical studies in collaboration with other laboratories or organisations.

The Aquatic Animals Commission assesses all applications for Reference Laboratory status and advises the OIE and the International Committee on their suitability. All Reference Laboratories are required to submit an Annual Report on their activities to the OIE, and these are collated and published by the OIE each year. The Commission assesses the reports of all the laboratories designated for one or more aquatic animal diseases, and considers whether there is sufficient evidence of relevant international activity to justify the continuation of their OIE Reference Laboratory status. Currently there are 36 Reference Laboratories with 23 Designated Experts for the aquatic animal diseases included in the *Aquatic Code* and *Aquatic Manual*.

OIE Collaborating Centres are centres of expertise in a specific designated sphere of competence relating to the management of general questions on animal health issues (for example epidemiology, risk analysis, etc.). In their designated field of competence, they must provide their expertise internationally. There is currently only one Collaborating Centre for aquatic animal diseases. The Centre's role is the provision of information on aquatic animal diseases through maintenance of an up-to-date on-line database of the official (reported to OIE) and non-official (published in scientific journals) information on the occurrence of the individual diseases. This is made accessible through the Centre's web site (www.collabcen.net) and the Commission's web pages on the OIE web site.

Dissemination of Information via the Commission's Web Pages

The Aquatic Animals Commission strongly supports the principle of openness and transparency in its activities, and provides an opportunity for anyone to follow its work through the Commission's pages on the OIE web site. The pages provide easy access to information on the terms of reference of the Commission, its current members, dates of Commission meetings, Commission meeting reports with proposed changes to existing OIE standards and any new standards, Ad hoc Group reports, the entire contents of the current editions of the *Aquatic Code* and *Aquatic Manual*, and the latest disease occurrence reports to the OIE by Member Countries. Information is provided on all the diseases covered by the *Aquatic Code* and *Aquatic Manual* and details are also provided for the OIE Reference Laboratories and Collaborating Centre for the aquatic animal diseases and the OIE Designated Experts. There is a link to the Collaborating Centre's web site and the International Database on Aquatic Animal Diseases that provides details of the recorded hosts and geographical distributions for all the diseases currently and previously listed by the OIE. On the home page, a 'focus on' section currently provides links to examples of national import risk analyses for fish diseases and national aquatic animal disease contingency plans.

The Commission's pages are being continually updated; the home page can be reached directly via http://www.oie.int/aac/eng/en_fdc.htm.

Conclusion

Due to the rapid global growth in aquaculture and the serious disease risks associated with the rising international trade in live aquatic animals and their products, OIE activities in the field of aquatic animal health are of increasing importance. The OIE Aquatic Animals Commission has been continuously developing the standards in the *Aquatic Code* and the *Aquatic Manual* to better meet the needs for preventing the spread of fish, mollusc and crustacean diseases through international trade, and to increase knowledge and awareness of their occurrence and distribution. Consideration is currently being given to adding amphibian diseases to the remit of the Commission. The work of the Commission is helped considerably by the specialist contributions of the Designated Experts in the OIE Reference Laboratories and its various Ad hoc Groups. All the activities and proposals of the Commission are transparent and can be easily viewed on its pages of the OIE web site. This dissemination is supplemented by Commission members representing the OIE and giving presentations on its work at international scientific conferences and at the conferences of the OIE Regional Commissions.

Acknowledgments

The Commission is supported at all its meetings by staff of the OIE Central Bureau, particularly Francesco Berlingieri of the International Trade Department and Sara Linnane of the Scientific and Technical Department. Their help in seeing that all the paperwork is available, requests for other Bureau staff to attend for specific items are met, and that the meeting report is completed to the required standard, is highly appreciated by members of the Commission. Don Lightner is a regular participant at our meetings, providing invaluable expertise in crustacean diseases, and we are often joined by Rohana Subasinghe of FAO, Rome, who ensures that Commission is kept appraised of aquaculture health circumstances in developing countries. We are all driven forward by Eva-Maria Bernoth, President of the Commission, who chairs our meetings with unrelenting determination, fortitude and good humour.

References

1. Office International des Epizooties (OIE): *Aquatic Animal Health Code*, 9th Ed., OIE, Paris, 2006.
2. Office International des Epizooties (OIE): *Manual of Diagnostic Tests for Aquatic Animals*, 5th Ed., OIE, Paris, 2006.

Dr. Barry J. Hill, Centre for Environment, Fisheries and Aquaculture Science, Barrack Road, Weymouth DT4 8UB, UK
E-mail: barry.hill@cefas.co.uk

Lombard M, Dodet B, the OIE Scientific & Technical Department (eds): First International Conference of the OIE Reference Laboratories and Collaborating Centres. Dev Biol (Basel). Basel, Karger, 2007, vol 128, pp 21-26.

The International Atomic Energy Agency (IAEA) Experience in the Development of Validation Systems Leading to the OIE Certification of Diagnostic Assays

H. Unger, J. Crowther

International Atomic Energy Agency, Vienna, Austria

Key words: IAEA, diagnostic assay, validation system, OIE certification

Summary: For the past 15 years, the International Atomic Energy Agency (IAEA) has had a major interest in helping its member states to diagnose livestock diseases through the development and validation of diagnostic tests for trans-boundary animal diseases. In November, 2007, an IAEA consultants meeting attended by experts, OIE members and producers reviewed the OIE process for the recognition of tests to date. Conclusions concerning the application procedure were tabled by the general meeting and then specific producer recommendations were made. Both are shown below, along with conclusions regarding referencing in general. Some of the points brought up suggest major changes in the idea of trade prescription and test status.

IAEA Activities

For the past 15 years, the International Atomic Energy Agency (IAEA) has had a major interest in helping its member states to diagnose livestock diseases through the development and validation of diagnostic tests for trans-boundary animal diseases. In this context, difficult issues of quality control have been addressed in a sustained way, and guidelines have been developed for good practice. Throughout this process, the IAEA has worked under the mandate of the OIE as the body responsible for producing guidelines and establishing principles of good practice. These have in turn been ratified by chief veterinary officers within the context of solving trade related disease problems.

The IAEA also fosters links with commercial producers since trends in funding are changing and support for extended research by governmental organisations is disappearing. The IAEA assumes that, in the long term, only commercial companies will be able to maintain a constant supply of quality controlled products such as diagnostic kits, and the OIE now accepts producer representation at IAEA and other meetings.

The consideration of how tests can be assessed, or "validated", is supported by the IAEA through their consultancies and experts and this has led to the addition of a Chapter in the *Manual of Diagnostic Tests and Vaccines for Terrestrial Animals.* To extend good practice in the area of validation, the IAEA organised key meetings in Vienna on *OIE Validation and Certification of Diagnostic Assays for Infectious Animal Diseases* in 2002 and 2003, where members of the OIE and FAO as well as world experts and producers met. The main finding was that there was a need to define tests/assays in terms of their intended use or fitness for purpose. Furthermore, it was found that there was a need to better define the criteria for validation and produce a guideline for good practice as a clear process in template form.

Subsequently, IAEA staff, working with a consultant, provided web based application forms for the certification of OIE tests based on fitness for purpose and validation guidelines agreed upon by consultants at previous meetings. The forms are the basis of the current certification pathway as shown on the OIE web pages. The IAEA initially recommended that their initiatives be reviewed before being used by the OIE, and offered a full, web-page-based, traceable, controlled-system version. Although this was not taken up by the OIE, it is available and could be readily activated.

The Most Recent Meeting on Validation

An IAEA consultants meeting in November, 2006, attended by experts, OIE members and producers reviewed the OIE process for the recognition of tests to date. Conclusions concerning the application procedure were tabled by the general meeting and then specific producer recommendations were made. Both are shown below, along with conclusions regarding referencing in general. Some of the points brought up suggest major changes in the idea of trade prescription and test status.

- The validation template is of a high standard and extremely useful as a guideline for developing tests used in the diagnosis and study of diseases.
- An OIE standard should be written for requirements and procedures for test certification.
- There is a need to review and expand the definitions of the terms used in the Web form and provide applicants with better guidance.

- The cost of registration with the OIE is too high.
- There is confusion over exactly what OIE registration means. Producers believed that a single approved registration process for all countries would be a great advantage to relieve them of the process of registering kits in many separate countries. Whether the OIE should try to establish this accepted registration process was discussed.
- The areas of trade prescribed tests and alternative tests were discussed and it was concluded that these terms now complicate validation assessment for countries and that they might be reassessed or abandoned and assessments made only in terms of fitness for purpose criteria.
- There was debate over the need for referencing in general and more specifically, standards for serological and molecular tests used for diagnosis and surveillance of livestock and human diseases. There can be no doubt that a critical factor in obtaining valid data from any study is the issue of quality of work and quality assessment and control. This can be exemplified by data from kits for the diagnosis of diseases that are affected by the robustness and formulation of the kit; the skills of the operators; the environmental conditions; the variable day to day and month to month performance of a test both within the same laboratory and between laboratories; data processing and misconceptions about the fitness for purpose of any set of reagents. This situation is further complicated by the lack of consistency in terminology; the assumptions made about status of laboratories; poor training of staff; lack of coordination by a variety of funding agencies and the existing rules on testing and emphasis on trade issues. Valid data based on defined sampling strategies was considered to be vital.
- It was agreed that central to the quality issue is the cost of production, gathering, characterization, storage and distribution of standard preparations, both for the continual assessment and the development of kits. Supply of well planned targeted reference materials is essential and at present weakly supported.
- It was concluded that decisions about standards for any situation are needed and should be made with expert advice, applied research and harmonization of efforts, particularly since funding for such efforts is woefully poor.
- A Portal Disease Information Exchange was also discussed. This could be a Web based, form driven (information has to fit into an established, easy to fill in template) system. Three data types for three areas were conceived: a) definitive data from validated assays; b) less definitive but extensive data from tests; and c) results where conditions are defined but only on a few samples. The portal could act as an active, continuously updated repository for data that can be viewed by anyone. The accumulation of data would

allow for statistics to be applied to a larger data set, thus improving validation and allowing a different stage of validation to be achieved though increasing confidence in the data. Management of the portal is required to: a) ensure correct input of data as well as to assign which "room" the data goes to; and b) to encourage better practice where data does not qualify for inclusion e.g. where internal quality control (IQC) is ignored. The IAEA will support this and hope that it can be successfully given over to the OIE in the future.

- The presence of experts from the human health fields was thought to be very useful in assessing the differences in the processes involved.

Below is a separate set of recommendations that was produced by the representatives of the European Association of Veterinary Diagnostics Manufacturers (Association européenne des fabricants de réactifs vétérinaires – AEFRV).

Recommendations of veterinary diagnostics manufacturers on the OIE registration procedure following the discussions at the expert's meeting at the IAEA in Vienna on November 21 to 24, 2006

Statement:
All AEFRV members agree on the usefulness of a unique, internationally recognized registration procedure (or international standards).

Recommendation Nr. 1: The recognition of the OIE certification procedure as a registration procedure

This is probably the most important requirement for the OIE procedure to be successful. Going through a certification procedure that is costly in time and money, but which does not take away local or national commercial barriers is considered useless for our commercial activity.

The OIE certification procedure needs to be "recognized" by the member countries that actively regulate Veterinary Diagnostics today as a registration procedure. The recognition means that registration by a national procedure will no longer be required. We, however, think it is reasonable that, in certain cases, additional data specific to a geographical region might be required as a complement to the registration. However, a correctly filled and validated template should include these data already ("fitness for purpose"). Hence the additionally required data should only concern the specific situation of a country.

Recommendation Nr. 2: Prescribed test, alternative test, and OIE registered test

The status of an OIE certified product versus a prescribed or alternative test described in the manual, needs to be revised and cleared up. The manual uses terms of prescribed test and alternative test (some of them are outdated) and the value of an OIE registration by the recent procedure is not specified. The term "prescribed" has a strong negative impact on the acceptance of use of other equivalent or superior certified kits.

A fit for purpose certified kit should become a "recommended" kit in the future. Veterinary diagnostics manufacturers understand that "old prescribed" tests should remain as "alternative" tests for countries which might have some difficulties accessing the recommended commercial kits.

Recommendation Nr. 3: Unfair competition from public and/or OIE reference laboratories; Conflict of interest

The unfair competitive position of some public laboratories, sometimes also OIE reference laboratories, which are selling their "prescribed" tests or kits or components regardless of their quality in comparison to commercially available kits, is considered as a big obstacle for veterinary diagnostics manufacturers. It would improve the image of the OIE to tackle the conflict of interest situation of some OIE reference laboratories. For example: the Veterinary Laboratories Agency (VLA) and Panaftosa have commercial activities by selling their "prescribed" test regardless of the fact that the quality very often is inferior. These laboratories control the market and make the market inaccessible to competitors. The manufacturers of veterinary diagnostics think that the OIE certification should also be a tool to clear up the unfair competitive position of these laboratories. A two-door access, one costly and seriously certified, next to another free of charge simplified without certification, is just unacceptable and does not serve animal health in general.

Recommendation Nr. 4: The cost of the procedure needs to be revised

The actual procedure costs 9,000 Euros plus an annual sales tax of 0.1%. Given the fact that the commercial activity of veterinary diagnostics is very different from human diagnostics, a substantial price revision is needed. The veterinary diagnostic market is a highly segmented market in terms of products. Most of the diagnostic manufacturers have more than 50 products on their catalogue.

The diagnostics manufacturers estimate that 2,500 to 3,500 Euros is a more reasonable fee, which will still cover the compensation of experts who need to be paid for their efforts in evaluating the dossiers (2 experts at 800€, and one chair expert at 1000€). Diagnostic manufacturers are unfavourable to the payment of an annual sales tax as it does not correspond to any service.

Recommendation Nr. 5: The procedure is considered too long and too complete

We recommend the suppression of Stage 4 validation. Stage 4 can only be reached if a kit has been on the market for a few years; hence it is impossible to fulfill this criterion for a new kit. In some cases, Stage 1 and 2 could already be sufficient.

Recommendation Nr. 6: Reference panels

Laboratories, which are running reference methods with certified kits, as well as manufacturers of veterinary diagnostic kits need access to international and national panel sera. Manufacturers are ready to pay a reasonable price to have access to these sera.

Considering that worldwide access would be preferable, this has to be handled by a worldwide-recognized organization, such as the OIE. Financing remains an open and important issue.

In essence, procedures to support a constant supply of validated diagnostic reagents in kit form and their international recognition or acceptance are progressing. The IAEA will continue to support these advances and at the same time evaluate the accepted procedures within the framework of their Technical Cooperation projects and Coordinated Research projects worldwide.

Dr. H. Unger, International Atomic Energy Agency, Vienna Austria
E-mail: h.unger@iaea.org

Lombard M, Dodet B, the OIE Scientific & Technical Department (eds): First International Conference of the OIE Reference Laboratories and Collaborating Centres. Dev Biol (Basel). Basel, Karger, 2007, vol 128, pp 27-35.

Development of a Framework for International Certification by the OIE of Diagnostic Tests Validated as Fit for Purpose

P. Wright[1], *S. Edwards*[2], *A. Diallo*[3], *R. Jacobson*[4]

1 Canadian Food Inspection Agency (NCFAD), Winnipeg, Manitoba, Canada
2 Veterinary Laboratories Agency (Weybridge), New Haw, Addlestone, Surrey, UK
3 International Atomic Energy Agency, Vienna, Austria
4 Eugene, USA

Key words: Infectious diseases, diagnostic tests, validation, certification, OIE Registry

Summary: Historically, the OIE has focussed on test methods applicable to trade and the international movement of animals and animal products. With its expanding role as the World Organisation for Animal Health, the OIE has recognised the need to evaluate test methods relative to specific diagnostic applications other than trade. In collaboration with its international partners, the OIE solicited input from experts through consultants meetings on the development of guidelines for validation and certification of diagnostic assays for infectious animal diseases. Recommendations from the first meeting were formally adopted and have subsequently been acted upon by the OIE. A validation template has been developed that specifically requires a test to be fit or suited for its intended purpose (e.g. as a screening or a confirmatory test). This is a key criterion for validation. The template incorporates four distinct stages of validation, each of which has bearing on the evaluation of fitness for purpose. The OIE has just recently created a registry for diagnostic tests that fulfil these validation requirements. Assay developers are invited to submit validation dossiers to the OIE for evaluation by a panel of experts. Recognising that validation is an incremental process, tests methods achieving at least the first stages of validation may be provisionally accepted. To provide additional confidence in assay performance, the OIE, through its network of Reference Laboratories, has embarked on the development of evaluation panels. These panels would contain specially selected test samples that would assist in verifying fitness for purpose.

Background

The OIE's *Terrestrial Animal Health Code* [1] describes international trade standards for terrestrial animals and their products. Within these standards, reference is made to 'prescribed' test methods. These are the assays required by the Code for the testing of animals before they are moved internationally. Prescribed tests may be found in the OIE's *Manual of Diagnostic Tests and Vaccines for Terrestrial Animals* [2]. The *Manual* is the international standard for diagnostic assays and is recognised as such in the Sanitary and Phytosanitary Agreement of the World Trade Organisation. Therefore, the *Code* and the *Manual* are companion standards, so much so, that the disease chapters in the *Manual* are organised in the same fashion as in the *Code* for ease of cross-reference. In addition to prescribed tests, the *Manual* also describes 'alternative' tests. These assays are suitable for the diagnosis of disease within a local setting, and can be used in the import/export of animals but only after bilateral agreement. A compendium of both prescribed and alternative tests, organised by disease, may be found in the *Manual* under the heading 'List of Tests for International Trade'.

The OIE's Biological Standards Commission, which is an elected body, oversees the production of the *Manual* and is responsible for establishing or approving test methods for the diagnosis of disease in mammals, birds and bees. As indicated above, application to international trade has been at the centre of the approval process. In addition to the many disease-specific chapters, the *Manual* begins with ten introductory chapters that cover a broad range of subject areas of general interest to infectious disease diagnostic laboratories. Two of these introductory chapters deal specifically with test method development and validation. The original chapter covers immunologic assays [3] and a more recent addition targets genomic assays [4]. Both chapters summarise the essential principles of assay development and validation that have been described in more detail by Jacobson [5]. These introductory chapters are intended to provide the scientific basis for validating assays as well as background information and guidance toward achieving a validated assay.

The OIE *Standard for Management and Technical Requirements for Laboratories Conducting Tests for Infectious Animal Diseases* [6] is a specific interpretation of the more generally stated requirements of the ISO/IEC 17025:1999 international quality standard for testing laboratories. The OIE Quality Standard clearly states that test methods and related procedures must be appropriate for specific diagnostic applications in order for the test results to be of any relevance. In other words, the assay must be 'fit for purpose'. The Quality Standard further states that in order for a test method to be considered appropriate, it must be properly validated and that this validation must respect the principles outlined in the validation chapters of the *Manual.*

The test methods currently described in the *Manual* are considered to be standard methods. Many of these test methods are 'classic' in that they have been used in diagnostic laboratories for many decades and their performance characteristics, including limitations, are well known. These assays, while still useful, will remain *status quo* in terms of reagents and format until replaced with newer technologies. The introduction of new test methods into the *Manual* has for the most part been initiated by test developers in national research establishments involved in regulatory diagnostic services. Prior to the addition of any new test method to a chapter, the developers are requested by the Biological Standards Commission to submit validation data for review. However, both the submission requirements and the peer review process have been rather informal. Based on reviewer's comments, the Commission then considers the addition of the method to the chapter and, if appropriate, its recommendation to the International Committee of the OIE for adoption as a prescribed or alternative test for trade. The above process has precluded submissions directly from the commercial sector.

In its role as the World Organisation for Animal Health, the OIE's scope of activities has broadened considerably over the past decade. In great part, this has been driven by the global impact of diseases such as bovine spongiform encephalopathy, foot-and-mouth disease and highly pathogenic avian influenza. So too, the need for appropriate test methods has expanded beyond trade applications alone. Although the *Manual* sets out the general principles of test validation, it was felt by the OIE and its international partners that there was a need for a more prescriptive framework for test developers to follow in the validation process. It was also recognised that the design of the validation studies should reflect the intended application of the assay.

First Consultants' Meeting

The Joint Division of the Food and Agriculture Organization of the United Nations (FAO) and of the International Atomic Energy Agency (IAEA) in Vienna, Austria, manages a large programme on 'Nuclear and Related Techniques in Food and Agriculture'. Animal health is a key component of the Animal Production and Health Sub-programme and the IAEA's Seibersdorf Laboratory is an OIE Collaborating Centre for ELISA and Molecular Techniques in Animal Disease Diagnosis. The Joint Division hosted two expert consultants meetings on the subject of 'OIE Guidelines for Validation and Certification of Diagnostic Assays for Infectious Animal Diseases'.

The first consultants' meeting was convened in Vienna in November, 2002. In attendance were programme experts from international organisations, including FAO, the IAEA, the OIE, World Health Organisation (WHO) and the Pan-American Health Organization. In addition, diagnostic experts from the UK, Canada, the USA, Australia, Denmark, and Sweden were invited. Three principle subject areas were addressed in detail:

1. Diagnostic test applications in animal health;
2. Validation and performance characteristics; and
3. Assay recognition or certification.

Given that animal disease management is required for economic, public health and environmental reasons, that risk assessment is a key component of this management and that risk assessment is highly dependent on the appropriate evaluation of animals and their products, it was agreed that fitness for purpose was integral to assay development and validation. These purposes or applications can be classified into six broad categories:

1. Demonstration of population 'freedom' from infection (prevalence apparently zero, where the term apparent 'freedom' recognises that absolute proof of freedom from infection in populations is not possible)
   a. 'free' with vaccination
   b. historical 'freedom'
   c. re-establishment of 'freedom' post-outbreak
2. Certification of freedom from infection or agent in individual animals or products for trade purposes
3. Eradication of infection from defined populations
4. Confirmatory diagnosis of clinical cases
5. Estimation of prevalence of infection to facilitate risk analysis (surveys, classification of herd health status, implementation of disease control measures)
6. Determination of immune status in individual animals or populations (post-vaccination)

It was agreed that the key data required to evaluate an assay comprised the following parameters:

1. Test type (detection of antibodies, infectious agents or their components);
2. Analytical sensitivity and specificity
3. Diagnostic sensitivity and specificity
4. Repeatability (within the laboratory)
5. Reproducibility (between laboratories)
6. Quality assurance capability
7. Throughput capacity in the laboratory
8. Turn-around-time of test
9. Technical sophistication (skills and equipment needed)
10. Interpretation skill required

In addition to these assay performance characteristics, factors such as the epidemiological parameters of the disease and the host population, sampling strategy and the structure and capability of the field veterinary services all have an impact on the deployment and application of a given test and must be taken into consideration when designing a disease control or testing programme.

With respect to a process for recognition or certification of assays, the participants proposed the following guidelines:

1. Assay developers should apply standard template requirements towards validation of a new test.
2. The total validation package should be evaluated by other laboratories (when they have not been involved in the original validation).
3. Evaluating laboratories must have established records in working with assays for the disease in question (at least one OIE Reference Laboratory if possible).
4. The template, with supporting documents, should be submitted to the OIE for evaluation.
5. The OIE should accept the assay after a positive and independent peer review of results.
6. The OIE should provide an independent opinion on the purpose(s) for which the assay is deemed to be fit at the time of the OIE evaluation. Any subsequent changes need re-evaluation and demonstration of equivalency or improvement.

Based on the above, the consultants' group drafted the following recommendations to the OIE:

1. The OIE should adopt a process for evaluation of diagnostic tests for specific purposes (based on the list above).
2. A standard validation template should be created to provide guidance to assay developers and to facilitate third party evaluation.
3. The template should support incremental stages of validation.
4. A registration process and a registry of certified test methods should be created and managed by the OIE.
5. OIE Reference Laboratories should establish characterised serum/sample collections to provide analytical references, evaluation panels and proficiency panels in support of the validation process.

A complete report of the first consultants' meeting was submitted to the OIE and was reviewed by the Biological Standards Commission at their biannual meeting in Paris in January, 2003.

OIE Resolution

At the 71st General Session of the OIE in Paris in May, 2003, the International Committee adopted Resolution XXIX which states that:

1. Fitness for purpose should be used as a criterion for validation
2. The Director General should make provisions to establish a registry for assays with levels of validation specified
3. OIE Reference Laboratories should be intimately involved with validation efforts
4. OIE Reference Laboratories should establish serum/sample reference collections

to be used for validation in line with their mandates

5. The Director General be given the mandate to review, in close consultation with the Standards Commission, the procedures involved in the timely approval of assays; this may entail the creation of a specific Ad hoc group comprising relevant experts to evaluate any submissions received
6. The Director General be authorised to recover if necessary, any costs incurred in the process of validating such assays

The above resolution gave the Director General of the OIE a clear mandate to proceed with the establishment of a new process for the evaluation and certification of diagnostic tests.

Second Consultants' Meeting

The second consultants' meeting was convened by both the Joint FAO/IAEA Division and the OIE in Vienna in December, 2003. The same experts were invited that participated in the first meeting. In addition, experts in the validation of assays in the absence of gold standards were invited from Denmark and Belgium. As the OIE resolution no longer places restrictions on who may submit validation dossiers, major stakeholders from the commercial sector were also invited. Three principle tasks were assigned to working groups selected from amongst the participants:

1. Finalise a prototype validation template
2. Propose an operating procedure for dossier submission
3. Develop guidelines for evaluation panels.

A detailed prototype template was presented by the working group tasked with this assignment. The prototype was agreed on by all of the participants. The key coponents of the prototype are listed below:

1. Background information
   a. Test method (functional description)
   b. Intended purpose(s) of test
   c. Applicant
   d. Scientific contact
   e. Accreditation status of developing laboratory
   f. Intellectual property considerations
2. Test method
   a. Protocol (detailed description of method and reagents)
   b. Kit configuration (if commercial)
3. Validation – stage I (analytical)
   a. Calibration data
   b. Repeatability

   c. Analytical specificity
   d. Analytical sensitivity

4. Validation – stage II (diagnostic)
   a. Reference animals (description)
   b. Threshold (cut-off) determination
   c. Diagnostic performance estimates
      i. Conventional estimates of diagnostic specificity and sensitivity using defined reference animals,
      ii. Mathematical estimates of diagnostic specificity and sensitivity using 'no gold standard' models, and
      iii. Comparison between currently used test(s) and proposed new test;

5. Validation – stage III (reproducibility)
   a. Laboratory selection criteria
   b. Evaluation panel (description)
   c. Reproducibility analysis

6. Validation – stage IV (current use)
   a. Laboratories currently using test method
   b. Assay application (purpose)
   c. International reference reagents (availability of calibration reagents)
   d. Inter-laboratory testing programmes (involving assay)
   e. International recognition (list of agencies/Organisations endorsing as an 'official' test method)

The second working group developed guidelines for an operating procedure for submission of dossiers for potential inclusion in the OIE registry. The principle essential features of this operating protocol included:

1. Purpose and aims of registry
2. Eligibility of submitters
3. Review panel selection and composition
4. Administrative responsibilities (OIE)
5. Guiding principles for review procedure
6. Conflict resolution
7. Registration as fit for purpose
8. Change control (affecting assay performance once registered)
9. Registration maintenance and renewal
10. Registration of current 'standard' methods
11. Registration fee structure
12. Glossary of terms and definitions (to assist developers and reviewers)

The third working group developed proposed recommendations for creation and use of evaluation panels. This latter task is somewhat multipurpose as evaluation panels potentially have more than one application in the assessment of both the test methods and the laboratories using them. This working group made recommendations related to the following:

1. Purpose and aims of evaluation panels
2. Confirmation of fitness for purpose assertions (as an adjunct to validation dossier)
3. Other applications (proficiency testing, serial release testing, etc,)
4. Composition and characterisation of evaluation panels (disease specific)
5. Responsibility for preparation and application (OIE Reference Laboratories)
6. Cost recovery of development and application of panels
7. Reporting and consequence of results

A complete report of the second consultants' meeting was submitted to the OIE and was reviewed by the Biological Standards Commission at their meeting in Paris in January, 2004.

Current Status

Over the period from 2004 to the present, a web version of the validation template as described above and application instructions have been developed and published on the OIE website (www.oie.int) and the registry is now open for submissions. A secretariat has been appointed within the OIE and a set of operating procedures established for the processing of applications. The process will be overseen by the Biological Standards Commission and it will appoint expert reviewers to assess validation dossiers. It is recognised that the template is designed around the validation of serological assays; however the same principles apply to other methods such as antigen detection and molecular methods. The template and procedures will be kept under review and will be updated and improved as necessary as in light of comments received from external sources.

Successful applicants will be allowed to use the OIE recognition as evidence of the validity of their test method for the stated purpose(s), and the register itself will be maintained on the OIE website. Tests on the register will be subject to periodic review and re-approval of their status.

Work has begun on the development of guidelines for the creation and application of evaluation panels. These guidelines will be distributed to OIE Reference Laboratories as part of their expanded role in the validation process.

Additional work is planned on the development of a template and guidelines for reviewers. This will ensure a uniform quality of review for all submissions. At the same time, the glossary of terms that currently accompanies the OIE Quality Standard will be reviewed and expanded to assist both test developers and reviewers. The glossary will be expanded to include more comprehensive epidemiological terms and new terms unique to molecular methods, as well as descriptive terms related to fitness for purpose.

The Biological Standards Commission is also developing links with statisticians and epidemiologists with special expertise in the application of latent class and extension of Bayesian theory to new models. This will add a valuable dimension to the design of test evaluation studies.

The OIE Registry and the supporting templates and guidelines have been developed to provide a formal mechanism for the certification and recognition of test methods as fit for purpose. However, all laboratories involved in test development or the application of diagnostic test methods are encouraged to use these templates and guidelines whether or not they intend to seek formal OIE recognition. In doing so, the quality of infectious disease diagnosis will be enhanced worldwide.

Acknowledgments

This paper was presented at the AFSSA, ISPAIA, IABS, USDA-APHIS, OIE International Conference on *"New Diagnostic Technology: Applications in Animal Health and Biologics Controls"*, Saint-Malo (France), 3-5 October 2005 and published in Dev Biol (Basel) 2006;126:43-51.

References

1. Office International des Epizooties (OIE): Terrestrial Animal Health Code, 14th Ed. OIE, Paris, 2005, 634 pp.
2. Office International des Epizooties (OIE): Manual of Diagnostic Tests and Vaccines for Terrestrial Animals, 5th Ed. OIE, Paris, 2004, 1178 pp.
3. Office International des Epizooties (OIE): Principles of validation of diagnostic assays for infectious diseases. In Manual of Diagnostic Tests and Vaccines for Terrestrial Animals, 5th Ed. OIE, Paris, 2004, 21-29.
4. Office International des Epizooties (OIE): Validation and quality control of polymerase chain reaction methods used for the diagnosis of infectious diseases. In Manual of Diagnostic Tests and Vaccines for Terrestrial Animals, 5th Ed. OIE, Paris, 2004, 30-36.
5. Jacobson, R.H.: Validation of serological assays for diagnosis of infectious diseases. Rev Sci Tech Off. Int. Epiz. 1998; 17 (2): 469-486.
6. Office International des Epizooties (OIE): OIE Quality Standard and Guidelines for Veterinary Laboratories, OIE, Paris, 2002, 63 pp.

Dr. Peter Wright, Canadian Food Inspection Agency, National Centre for Foreign Animal Disease, 1015 Arlington Street, Winnipeg, Manitoba, R3E 3M4, Canada.
E-mail: pwright@inspection.gc.ca

Lombard M, Dodet B, the OIE Scientific & Technical Department (eds): First International Conference of the OIE Reference Laboratories and Collaborating Centres. Dev Biol (Basel). Basel, Karger, 2007, vol 128, pp 37-39.

International Harmonisation

P. Dehaumont, C. Lambert

OIE Collaborating Centre on Veterinary Medicines, AFSSA-ANMV, Fougères, France

Key words: Veterinary medicinal products (VMPs), veterinary medicines, OIE Collaborating Centre, AFSSA-ANMV

Summary: Like several other sectors, animal and public health face new challenges arising from the worldwide globalisation phenomenon. To this end, the 4th strategic plan specifically addresses reinforcing the capacity of Regional Representations and the network of OIE Collaborating Centres and Reference Laboratories and implementing networking approaches. A good example of achievements in this field, and of overcoming constraints and limits, can be found in veterinary medicinal products (VMPs).

Like several other sectors, animal and public health face new challenges arising from the worldwide globalisation phenomenon. Towards this end, the OIE has recently refined its objectives, which can be summarised as follows:

- to ensure transparency with respect to the global animal disease and zoonosis situation;
- to collect, analyse, and disseminate scientific veterinary information;
- to develop and encourage international solidarity by providing expertise;
- to safeguard world trade of animal and animal products under its World Trade Organisation (WTO) mandate;
- to implement new mandates dedicated to food safety at the level of primary production;
- to support animal welfare;
- to support strengthening and capacity building of veterinary services, now considered as a public good.

These general OIE objectives are key drivers for ongoing implementation of the 4th strategic plan.

Along with OIE missions and objectives, Collaborating Centres (CC) and Reference Laboratories (RL) are key tools, providing expertise and assistance in their sphere of competence. For decades they have worked hard and they will have to enhance their responsibilities and duties in the future in order to meet new challenges and new goals. Consistent with the above mentioned elements, the 4th strategic plan specifically addresses reinforcing the capacity of Regional Representations and the network of OIE Collaborating Centres and Reference Laboratories and implementing networking approaches.

The tools and expertise developed by CC and RL are based on the elaboration of international standards, cooperation with other international organisations and technical assistance. Although major achievements have been reached during the past decades, there are still many constraints, a large portion of which are linked to the heavy workload associated with information dissemination, faced by the RL and CC.

A good example of achievements in this field, and of overcoming constraints and limits, can be found in veterinary medicinal products (VMPs). By designating the French Agency for Veterinary Medicines (AFSSA-ANMV) as an OIE Collaborating Centre for VMPs, the OIE highlighted the importance of VMPs for animal health, public health and the facilitation of trade. Currently, there is a growing demand for healthy animals receiving relevant preventive or curative treatments and animal food products that are free from residues. The main activities of AFSSA-ANMV are threefold: international harmonisation, expertise, and technical assistance. The AFSSA-ANMV is involved in international harmonisation through its participation in Codex alimentarius meetings and collaboration with WHO and FAO for risk assessment and risk management of VMPs. The AFSSA-ANMV also participates in the OIE Animal Production Food Safety Working Group (APFSWG).

The agency contributes to the International Cooperation on Harmonisation of Technical Requirements for Registration of Veterinary Medicinal Products (VICH), which was created under the auspices of the OIE more than ten years ago. VICH is an international program with the aim of establishing harmonised regulatory requirements for veterinary medicines registration.

The AFSSA-ANMV also participated in the elaboration of the procedure for diagnostic kit certification, which was adopted in 2004, and supports its implementation. The aim of the certification procedure is to help OIE member countries better protect animal and public health, taking into account the fact that this procedure could be applied as a national system model to facilitate worldwide harmonisation. As a collaborating centre, the AFSSA-ANMV also delivers its expertise in different areas such as antimicrobial resistance. During the past few years, the Agency contributed to the elaboration and adoption of five guidelines related to antimicrobial resistance. After two joint FAO-WHO-OIE consultations (Geneva, 2003, and Oslo, 2004) on non-human antimicrobial resistance, the OIE decided to build a list of veterinary critically important antimicrobial agents (VCIA). The AFSSA-ANMV is a member of the OIE Ad hoc group that compiled a list of VCIA based on the information supplied by OIE. The Ad hoc group report was presented at the 2006 general session and Resolution N° XXXIII asked the group to refine the list. The group suggested a new list in September 2006, to be presented at the next session. Considering the multidisciplinary aspect of the antimicrobial resistance issue, in 2007, the AFSSA-ANMV, as an OIE collaborating centre, will participate in the task force on risk management options created following a joint FAO-WHO-OIE expert consultation.

Finally, the AFSSA-ANMV also provides technical assistance to countries or regions to implement harmonised VMP regulations. For instance, it has been deeply involved in setting up the legislative corpus for the registration of VMPs in the West African Economic and Monetary Union (UEMOA), and in the implementation of good manufacturing practices in Vietnam.

At this stage, however, the limits of the system have been reached. Due to constraints on CC and RL resources it will be difficult to develop additional projects unless a new way of functioning is found. Thus, for the future, the way forward will be to keep on with existing activities, to strengthen support for the implementation of existing standards and to develop a wider scope of technical assistance and harmonisation through the identification of regional specialised centres in developing countries. New ways of functioning that would allow for additional projects might include: work-sharing, best practices exchanges, the development of new competences, networking for better dissemination of information and capacity building. A worldwide network of CCs and RLs should be developed to sustain these practices. It is recommended that a "twinning approach" be developed with the strong support of the OIE headquarters.

Dr. Patrick Dehaumont, AFSSA-ANMV, OIE Collaborating Centre on Veterinary Medicines, Agence nationale du médicament vétérinaire, B.P.203, La Haute Marche, Javené, 35302 Fougères Cedex, France
E-mail: p.dehaumont@anmv.afssa.fr

Lombard M, Dodet B, the OIE Scientific & Technical Department (eds): First International Conference of the OIE Reference Laboratories and Collaborating Centres. Dev Biol (Basel). Basel, Karger, 2007, vol 128, p 41.

Session II

Application of Laboratory Practices

Lombard M, Dodet B, the OIE Scientific & Technical Department (eds): First International Conference of the OIE Reference Laboratories and Collaborating Centres. Dev Biol (Basel). Basel, Karger, 2007, vol 128, pp 43-46.

The OIE and the World Association of Veterinary Laboratory Diagnosticians: A New Partnership

C. Carter

Professor, Epidemiology, Livestock Disease Diagnostic Center, College of Agriculture, College of Public Health, University of Kentucky, Lexington, KY, USA

Key words: WAVLD, OIE, veterinary diagnostic laboratory, animal health, zoonoses, food safety

Summary: The OIE and the WAVLD have many shared goals and aspirations and the Board of Directors of the WAVLD are very excited and proud of this new partnership. Although the WAVLD is a relatively small organization with limited financial means, it is hoped that our relationship with the OIE will allow us to grow and to possibly provide services beyond the biennial ISWAVLD educational symposia and informal consultations. To be sure, many countries are struggling to build a strong veterinary diagnostic laboratory infrastructure to serve their animal agricultural industries. War-torn countries like Iraq and Afghanistan are also in great need of help in this realm. Few initiatives have the potential to have such a huge and positive impact on the people and animals of a country like the implementation and operation of a sound veterinary diagnostic laboratory system. However, it will take considerable funds to provide the expert assistance that is necessary to help build such a system. The WFAH proposal is still under consideration by the WAVLD Board. Please contact us with your ideas on this and I will share them with the Board. The WAVLD web site is located at: www.wavld.org.

Introduction

The mission of the World Association of Veterinary Laboratory Diagnosticians (WAVLD) is to enhance animal and human health by helping countries provide the finest veterinary diagnostic laboratory testing services through the application of cutting-edge science and technology. This mission is met by disseminating the latest information relating to the diagnosis of animal diseases through outstanding educational symposia, promoting the organization of associations of veterinary laboratory diagnosticians, and providing consulting assistance when possible to countries wishing to build and operate state-of-the-art veterinary diagnostic laboratories. In February, 2006, a partnership was formed between the OIE and the WAVLD based on several shared goals.

WAVLD History

In 1976, Dr. Benjaman Jarra of Mexico visited the National Animal Disease Laboratory in Ames, to consult with Dr. Vaughn Seaton and others regarding a new Veterinary Research and Diagnostic Laboratory at Palo Alto, Mexico that was to be dedicated in 1977. After much discussion, they decided that an international organization was needed to provide planning guidance for countries wanting to establish a veterinary diagnostic laboratory service to support their animal agricultural industries. Planning then commenced for an organizational meeting of the WAVLD to coincide with the dedication of the new laboratory in Palo Alto.

The organizational WAVLD meeting was held in Guanajuato, Mexico in 1977 with several hundred in attendance. At the Mexico meeting, Dr. Hans Glattli lobbied to host the second meeting in Lucerne, Switzerland, in 1980. At the 1980 meeting in Switzerland, it was decided to name the new organization the *World Association of Veterinary Laboratory Diagnosticians*. Dr. Vaughn Seaton of Ames, Iowa, was installed as the first WAVLD President and remained in that position for six years. It was also decided that similar meetings would occur every three years and would be titled *"The International Symposium of the World Association of Veterinary Laboratory Diagnosticians"* or ISWAVLD. At the ISWAVLD meeting in College Station, Texas, USA in 1999, it was decided to make the meeting a biennial event.

The WAVLD constitution and bylaws were written and approved at the third ISWAVLD meeting in Ames, Iowa, USA in 1983. The purpose of the WAVLD was outlined as follows:

- Disseminate the latest information relating to the diagnosis of animal diseases through outstanding educational symposia;
- Facilitate the organization of associations of veterinary laboratory diagnosticians in all countries of the world;
- Provide consulting assistance to countries wishing to build and operate state-of-the-art veterinary diagnostic laboratories;
- Conduct other activities that may improve the health and welfare of man and animals throughout the world.

The events of the years following the 1983 meeting comprise a rich history of educational symposia and collaborative activities among the finest veterinary laboratory diagnosticians in the world. Dr. Konrad Eugster assumed the role of Secretary-Treasurer of WAVLD in 1992 and provided the strong leadership, brilliance and energy necessary to build attendance for future ISWAVLD meetings. Below is a list of the locations of the ISWAVLD meetings dating back to 1977 [1]:

- 2005, Montevideo, Uruguay
- 2003, Bangkok, Thailand
- 2001, Parma, Italy
- 1999, College Station, TX, USA
- 1996, Jerusalem, Israel
- 1994, Buenos Aires, Argentina
- 1992, Lyon, France
- 1989, Guelph, Ontario, Canada
- 1986, Amsterdam, Netherlands
- 1983, Ames, Iowa, USA
- 1980, Lucerne, Switzerland
- 1977, Guanajuato, Mexico

The next ISWAVLD meeting will be held in Melbourne, Australia, Nov 11-14, 2007. The Australian delegation is already building an outstanding scientific program for what will be a memorable symposium. Day two of the symposium will be the *World Organisation of Animal Health (OIE) Biotechnology Symposium.* The OIE focus will be *Applications of Biotechnology in the Study of Pathogens and Pathology of Animal Diseases.* They have already delivered a call for papers. More information can be found relating to the upcoming meeting at: www.wavld2007.com. Furthermore, the 2009 meeting will be held in Madrid, Spain.

OIE/WAVLD Collaboration

In February, 2006, a partnership was formed between the OIE and the WAVLD based on several shared goals and objectives [2]. The first of theses is advancing the use of diagnostic tests in the control of infectious animal diseases and zoonoses in the world. The second is helping to organize associations of veterinary laboratories and diagnosticians in all countries of the world. The third issue is the enhancement of the capability of Member Countries on the standardization and harmonization of the use of diagnostic techniques according with the OIE standards. The fourth is to disseminate the latest information relating to the diagnosis of animal diseases and zoonoses through outstanding continuing educational workshops, seminars, and international conferences. The fifth is to provide consulting assistance to Member Countries on the state of the art of diagnostic technology and accreditation for veterinary laboratories. The sixth is to provide advice on veterinary research related to diagnosis of animal diseases and zoonosis. Finally, we agreed to engage in other activities that may be conducted to improve the health and welfare of animals throughout the world.

Proposed *World Foundation for Animal Health*

At the ISWAVLD meeting in Montevideo, Uruguay in 2005, the WAVLD Board discussed the formation of a new organization called the World Foundation for Animal Health (WFAH) [3]. The WFAH would be a non-profit, fund-raising group with the express goal of offering expert consulting services to developing countries that wish to develop a strong veterinary diagnostic laboratory infrastructure. Funding for consultants would come from endowments established by various benefactors. To be certain, OIE diagnostic testing and ISO 17025 standards and guidelines would be the foundation of this work. It is hoped that these efforts would help to achieve optimum animal health through early and accurate laboratory diagnosis and the prevention and control of diseases common to animals and man (zoonoses), and the enhancement of food safety.

References

1. Proceedings of the ISWAVLD scientific symposium, years 1983-2005, some available through the WAVLD.
2. Letter from Dr. Bernard Vallat delivering signed copies of the cooperative agreement between the WAVLD and the OIE, 6 February 2006.
3. Minutes of the WAVLD Board of Directors, Montevideo, Uruguay, 19 November, 2005.

Pr. Craig Carter, Epidemiology, Livestock Disease Diagnostic Center, College of Agriculture, College of Public Health, University of Kentucky, Lexington, KY, USA
E-mail: Craig.Carter@uky.edu

Lombard M, Dodet B, the OIE Scientific & Technical Department (eds): First International Conference of the OIE Reference Laboratories and Collaborating Centres. Dev Biol (Basel). Basel, Karger, 2007, vol 128, pp 47-51.

The Changing World of Biocontainment and Biosecurity

M. Jeggo

OIE Collaborating Centre for New and Emerging Diseases, Australian Animal Health Laboratory (AAHL), Australia

Key words: Biocontainment, biosecurity, biosafety, veterinary laboratory, Australian Animal Health Laboratory

Summary: Since the 1980s, a number of key events have significantly altered our ideas on biosecurity and the role of biocontainment laboratories, such as the bovine spongiform encephalopathy (BSE) and severe acute respiratory syndrome (SARS) outbreaks and the anthrax episode of 2001 in the USA. This has resulted in the development of plans to build "high containment" facilities around the world and an array of new regulations at both national and international levels regarding the management of pathogens, the operation of high containment facilities, the use of genetically modified material, and the transportation of such agents and personnel security issues. Considering the cost, however, it is debatable whether every country needs to build its own high containment facility. The Australian Animal Health Laboratory (AAHL) provides an example of what might be considered best practice in biocontainment while considering regulations, cost and need.

INTRODUCTION

A number of key events have significantly altered both the environment and the risks associated with biosecurity and the role of biocontainment laboratories. Currently, in simplistic terms it is suggested that biocontainment refers to "managing the risks from the pathogens", whilst biosafety means "managing the risks to humans from the pathogens" and the broad term biosecurity now refers to "managing the risks associated with living things (e.g. infectious agents, invasive species, weeds, people).

So what has changed that has led to increased focus and enhanced perceived risks in this area? In the late 1980's many thought that infectious diseases were well under control, that the major global infectious diseases such and rinderpest and foot-and-mouth disease (FMD) were being eliminated from most parts of the globe, and that the risks from infectious diseases were rapidly diminishing. A number of events in the following 15 years, however, illustrated clearly that this was not the case. The outbreak of bovine spongiform encephalopathy (BSE or mad cow disease) was perhaps the first event that brought a rude awakening to the risks that a "new" infection could pose to both animals and humans. This was followed over the next few years with devastating outbreaks of FMD in Europe and elsewhere, with severe acute respiratory syndrome (SARS) in China, Hong Kong and Canada and more recently with the avian influenza (AI) pandemic. Notwithstanding these disease events, clearly the anthrax episode of 2001 in the USA changed our thinking profoundly, and brought new risks associated with terrorism and more specifically bioterrorism. The anthrax events in the USA brought a menacing focus to these areas with the formation of the United States Department of Homeland Security and the new concept of managing selected pathogenic agents through a bioterrorist risk-based process. Finally, and less threatening but equally important, the concept of genetically modified organisms and trans-genetic animals became realities and we saw an increasing focus on processes and regulations to manage the risks associated with these advances in science.

All of this has resulted in the development of plans to build "high containment" facilities around the world to better manage the risks associated with all or some of the above agents. A careful review of two of the world's oldest and largest high-containment facilities in the UK (Pirbright) and the USA (Plum Island) led to a realization that they were no longer viable and would both need complete rebuilding. All of this has been associated with an array of new regulations at both national and international levels around the management of the pathogens, the operation of high containment facilities, the use of genetically modified material, and the transportation of such agents and personnel security issues. Compliance monitoring in many of these areas is at best, in its infancy.

It is now a complex world that encompasses biosecurity and biocontainment. Are there things we should be doing differently or better?

The Australian Example

The Australian Animal Health Laboratory (AAHL) based in Geelong, in the State of Victoria, Australia is probably the world's most advanced high security laboratory. Opened in 1985, it is currently undergoing a 25A$ million upgrade which will be completed next year. It is worth examining this laboratory in considering what might be considered best practice in biocontainment and issues surrounding regulations, cost and need.

Who and what is regulated in Australia? The Australian/New Zealand Standards define the minimum requirements for safety in biological laboratories and are concerned principally with the safety of workers and the community. The Australian Quarantine Inspection Service (AQIS) controls the importation and subsequent management of any material to which quarantine restrictions apply.

More recently the Commonwealth Department of Health and Ageing (DOHA) has developed a series of regulations and procedures that govern the use of a group of selected pathogenic agents whilst the Office of the Gene Technology Regulator (OGTR) manages the regulation of genetically modified organisms.

The Australian Standards and DOHA use Risk Groupings in their determinations, whilst AQIS and OGTR specify Physical Containment Levels (PC 1-4). The Standards are advisory; AQIS and OGTR are compliance monitored whilst the new DOHA regulations will soon be. It is within this complex environment that the AAHL must operate.

Physical Bio-Containment and Biosafety at the AAHL

In design and operation, the AAHL applies a multiple systems approach that ensures that there is considerable redundancy such that a single point failure has no immediate biological or operational risk. Redundancy is an unfortunate word but as used, indicates that if one system fails at least one other will be operational and ensure the microbiological safety of the facility. There are a range of technical approaches utilized including increasing negative air pressure as the biological risk increases, the box within a box principle, the extensive use of air filtration (some 1,000 HEPA filters permanently in operation), solid waste disposal through continuous flow and batch treatment of liquid effluent as well as incineration of solid material. There are many highly complex systems of continuous monitoring (128 separate systems and over 4,000 sensors routinely recording) with permanent auditing and varying the location of all agents held.

Staff Management

It has been recognized for some time that the "insider threat" is significant for such facilities as AAHL and therefore considerable attention is paid to managing this risk. All staff are required to have a full security screening. Access throughout the facility is determined on a case by case basis with each access card individually tailored to the needs and skills/training of the staff member (or visitor). Since access to all locations is through the use of card readers it is possible at one point in time to determine with a close proximity, the exact location of each staff member. Access to a number of highly sensitive areas or those requiring specialized skills, is further restricted using key pads requiring a personalized 4 digit code. This double access control provides considerable added security and is a double safeguard against "gate tailing". Although in theory it would be possible for someone to follow in a card-carrying staff member without having to use a card, unless this was to happen throughout their sojourn around the facility they would soon become trapped and unable to go forwards or backwards! It should be noted that there are emergency overrides at all key points, but use of these alerts the monitoring staff.

Linked to all the above is an extensive staff training program that repeats on a routine basis and is a prerequisite to maintaining entry rights.

The use of the facility by visitors and temporary staff poses particular challenges to this system and for the most part relies on close (permanent) supervision of this category of staff. This is not always convenient and is particularly onerous out of normal working hours, but nevertheless is now strictly followed.

Five Stories of Technology

The building itself consists of five stories, the top two dealing with air handling, the middle floor with the science whilst the lower two floors deal with solid waste management and disposal. There are many unique design features in the buildings and the major upgrade has further advanced these to use cutting-edge technologies. Air handling has been upgraded to use variable speed fans to control air flow and maximize efficiency and energy usage. Similarly, advances in batch treatment systems have increased sewage management efficiency and reduced the need to over-utilize the less efficient continuous flow treatments.

A simple illustration of the redundancy principle is seen in the approach to standby electricity generation. Naturally, any such laboratory has a backup diesel or equivalent standby generator, but AAHL has not one, not two, but three of these which are continually maintained in readiness for use.

What is the Lesson?

Who would argue that AAHL operates at the level of the world's best practice for a high containment facility? Nevertheless it must be recognized that the underlying principle that governs these process and engineering practices must be that they are risk-based as they are incredibly costly to build, to operate and to comply with! Furthermore they should have continuous external oversight. In the AAHL case this is through the AAHL Security Advisory Group that comprises outside experts who report not to the Director of AAHL but one level above to ensure effective accountability.

Given the above mentioned cost, it is obvious that we should determine the real need for such facilities, that we should take a global approach and that we should maximize the use of such facilities. It makes little sense for every country to build one, and even less if the high cost of daily operation cannot be maintained.

The AAHL is currently under-utilized in terms of maximizing the floor space that can be used as laboratories in the "secure area", and a further 2,000 square meters of laboratory space is being brought into routine use to maximize the space available at AAHL. Already that space has been taken up by other research groups across Australia desperate to undertake work in a high containment laboratory.

There are currently plans in a number of countries to build new high containment facilities and both facilities in the UK and USA will be completely re-built. Do we really need all these and can we safely operate and maintain these once they become operational? Agreement should be reached on the various risks involved, and harmonize our approach, rules, and monitoring. It is therefore suggested that the OIE should undertake to:

- establish international laboratory standards for (veterinary) laboratory bio-containment;
- and bio-safety;
- establish standard operating procedures (SOPs) for bio-containment activities;
- develop an oversight and compliance monitoring system;
- drive the process of global partnerships and sharing of resources.

Dr. Martyn Jeggo, OIE Collaborating Centre for New and Emerging Diseases, Australian Animal Health Laboratory (AAHL), Australia
E-mail: marga.jeggo@csro.au

Lombard M, Dodet B, the OIE Scientific & Technical Department (eds): First International Conference of the OIE Reference Laboratories and Collaborating Centres. Dev Biol (Basel). Basel, Karger, 2007, vol 128, pp 53-56.

Quality Systems in Veterinary Diagnostics Laboratories

L.M. Branco de Freitas Maia

Ministerio da Agricultura, Pecuaria e Abasteciment, Brasil

Key words: Quality systems, veterinary diagnostic laboratories, ISO 17025:2005, ISO 9001:2000.

Summary: Quality assurance of services provided by veterinary diagnostics laboratories is a fundamental element promoted by international animal health organizations to establish trust, confidence and transparency needed for the trade of animals and their products at domestic and international levels. It requires, among other things, trained personnel, consistent and rigorous methodology, choice of suitable methods as well as appropriate calibration and traceability procedures. An important part of laboratory quality management is addressed by ISO/IEC 17025, which aims to facilitate cooperation among laboratories and their associated parties by assuring the generation of credible and consistent information derived from analytical results. Currently, according to OIE recommendation, veterinary diagnostics laboratories are only subject to voluntary compliance with standard ISO/IEC 17025; however, it is proposed here that OIE reference laboratories and collaboration centres strongly consider its adoption.

INTRODUCTION

Quality assurance of services provided by veterinary diagnostics laboratories is a fundamental element promoted by international animal health organizations to establish trust, confidence and transparency needed for the trade of animals and their products at domestic and international levels. The quality system supporting the required levels of confidence in animal diagnostics require, inter alia, trained personnel, consistent and rigorous methodology, choice of suitable methods as well as appropriate calibration and traceability procedures. The managing of these and other applicable requirements within a quality system is known as laboratory quality management.

This work examines and discusses relevant issues associated with the elementary components of quality management in veterinary diagnostics laboratories. It places particular emphasis on biosafety, worker's health, environmental concerns, and on the need for unrestricted commitment of all laboratory personnel to the pursuit of quality, which is a mainstream philosophy in the daily routine of the laboratory.

Development

The International Organization for Standardization (ISO)

The International Organization for Standardization (ISO) is a non-governmental initiative comprising 150 member countries, recognized worldwide for voluntary standardization. ISO aims at promoting standardization as a tool to facilitate international trade of services and products, fostering broader economical, technological and scientific cooperation and development.

The fundamental output of ISO is the publication of international standards produced by consensus and applicable to a wide array of processes. Prior to final approval and issuance, projects for future standards are submitted to a vote by member countries. Publication of an international standard requires approval of at least 75% of the voting members.

In the particular case of laboratories, the ISO/IEC standard identified by the number 17025 establishes General Requirements for the competence testing and calibration laboratories. The second revision of this standard (ISO 17025:2005) contains altered or added sections in order to include the general quality assurance procedures of standard ISO 9001:2000. As a result, laboratories which work with the current ISO 17025:2005 work according to ISO 9001:2001.The chief objective of standard ISO/IEC 17025 is to facilitate cooperation among laboratories and their associated parties (e.g., private clients, regulatory authorities and accreditation bodies) by assuring the generation of credible and consistent information derived from analytical results. Therefore, the ISO/IEC 17025:2005 standard should be the mainframe of a laboratory´s evidence of adoption of a broad quality management system - covered by the inclusion of ISO 9001:2000 directives in this second revision of the standard - and its competency.

However, compliance with standard ISO/IEC 17025 should not serve as a basis for laboratory certification. In addition, the safety of laboratory operations and other specific operative requirements are not covered by this standard.

Reference Laboratories and OIE Collaboration Centers

The OIE mission is stated as follows:

"To ensure transparency in the global animal disease and zoonosis situation;
to provide expertise and encourage international solidarity in the control of animal diseases;
within its mandate under the WTO SPS Agreement, to safeguard world trade by

publishing health standards for international trade in animals and animal products;
to improve the legal framework and resources of national Veterinary Services;
to provide a better guarantee of the safety of food of animal origin and to promote animal welfare through a science-based approach."

In this capacity, the OIE sets international sanitary rules with reference status within the WTO. These rules are elaborated by elected Specialized Commissions and by Working Groups composed of leading scientists in their fields. The majority of these scientists are associated with a network of 160 Collaboration Centers and Reference Laboratories that cooperate to achieve the OIE´s scientific objectives.

OIE Reference Laboratories, each represented by a specialist, are appointed to meet investigation needs of every key animal disease listed by the OIE. Besides their duties in methodology and method standardization, these laboratories also provide analytical services in case of trade and legal disputes.

Currently, according to recommendation, veterinary diagnostics laboratories are only subject to voluntary compliance with standard ISO/IEC 17025; however, this standard has been effectively internalized by the OIE as a basis for accreditation of laboratories performing diagnosis for infectious diseases in animals. Laboratories involved in such controls for international trade are particularly subject to this policy. As discussed previously, the complying laboratories both demonstrate that a broad and general quality system is in place, and show evidence of technical competency.

With the intent to offer a practical view of the difference between providing evidence of an existing general quality system and evidence of technical competency, one may consider questions that would be asked by an auditor in each case.

The following examples, as described by A. L. Wiegers, are useful to illustrate the concepts discussed.

In the first example, an ISO 9001 auditor would ask: *"Have you defined your policies and procedures? Are they documented according to a standard? Do you actually follow them as documented?*" Thus, this audit is designed to show evidence that the general quality system under examination meets the standardized requirements.

For the second example, an auditor for technical competency would ask: *"Have you defined and validated your analytical procedures, and are they documented following the standard's requirements? Do you actually follow them as documented? Can you show evidence that these procedures yield credible and consistent results? Do you understand the scientific principles underlying the procedures? Can you predict and overcome technical problems that could arise? Is your equipment suitable for the task? Do you have trained personnel, and how are they trained? Have you calculated the uncertainties associated with your method or do you know the uncertainties associated with it?"* The concept of technical competency, as one can see with this example, refers to the validity of laboratory procedures, and the understanding of them by involved laboratory personnel, so that the underlying science is put to optimum practical use. It is also related directly to the ability of the laboratory to be organized and proactive towards technical excellence.

Given the clear benefits resulting from compliance with standard ISO/IEC 17025:2005, it is hereby proposed that OIE reference laboratories and collaborating centres strongly consider its adoption. In its turn, promotion of compliance to standard ISO/IEC 17025:2005 is best led by the OIE via *in situ* verification of reference laboratory quality systems, associated with appropriate inter-laboratory proficiency rounds. This would certainly broaden their already established, unequivocal scientific competency to the boundaries of demonstrated analytical excellence.

References

1. Manual de las Pruebas de Diagnóstico y de las Vacunas para los Animales Terrestres, quinta edición, OIE, 2004.
2. Norma de calidad y directrices de la OIE para los laboratorios veterinarios: enfermedades infecciosas, OIE, 2002.
3. Ann L. Wiegers. The age of competence: an update on the international laboratory accreditation scene for veterinary testing laboratories, Special Article: J Vet Diagn Invest 2002;14:89-96.
4. International Organization for Standardization (ISO): ISO / IEC International Standard 17025. General Requirements for the competence of testing and calibration laboratories. 2005;1-26.

Dr. Lucia Maria Branco de Freitas Maia, Ministerio da Agricultura, Pecuaria e Abasteciment, Esplanada dos Ministérios BI D – anexo B sala 432, Brasilia, DF 70043-900, Brasil
E-mail: luciamaia@agricultura.gov.br

Lombard M, Dodet B, the OIE Scientific & Technical Department (eds): First International Conference of the OIE Reference Laboratories and Collaborating Centres. Dev Biol (Basel). Basel, Karger, 2007, vol 128, p 57.

Session III

Current Status and Approval of Laboratory Networking

Lombard M, Dodet B, the OIE Scientific & Technical Department (eds): First International Conference of the OIE Reference Laboratories and Collaborating Centres. Dev Biol (Basel). Basel, Karger, 2007, vol 128, pp 59-68.

Regulatory Constraints for the Transport of Samples and Compliance with the World Organisation for Animal Health (OIE) Standards for Biosecurity and Biocontainment

J.E. Pearson

World Organisation for Animal Health (OIE), Paris, France

Key Words: Infectious substances, dangerous goods, biocontainment, biosecurity, United Nations Model Regulations

Summary: The International Regulations for transport of infectious substances, including diagnostic specimens, are based on the United Nations Model Regulations and are the standard for transport by all means of transportation including air transport; the International Air Transport Association (IATA) regulation specifically addresses air shipment. In 2005 and 2006 there were major improvements in the procedures for shipping infectious substances. These substances are divided into Category A, which includes primarily cultures of the more pathogenic agents and Category B, all the other substances. Category A shipments must have a Dangerous Goods Certificate and meet other requirements. Category B shipments, which include most diagnostic tissue specimens, do not. These regulations specifically exempt certain substances, including those that have been neutralized or inactivated to destroy pathogens and samples from "normal" animals. The packaging requirements help insure that biocontainment is maintained during shipment to protect the shipper and the environment. The packaging requirements and the shipping procedures provide a chain of custody and assist in supporting biosecurity. The more stringent Category A requirements provide increased biocontainment and biosecurity safeguards for these potentially more dangerous substances. In addition, National requirements, such as import permits and the US select agent requirements, provide an added measure of biocontainment and biosecurity.

INTRODUCTION

Regulations for the shipment of dangerous goods, which can include diagnostic specimens, are developed to provide biocontainment measures that protect shipping personnel and the environment. The Model Regulations for the transport of dangerous goods are developed by the United Nations Sub-Committee of Experts on the Transport of Dangerous Goods (UN SCETDG) and are applicable to all modes of transportation. These regulations outline the requirements for the international transport of biological materials, which include diagnostic samples, reagents, cultures, and reference material. The 13th revision of the Model Regulations made major changes in the shipping requirements for transport of dangerous goods, including infectious substances, and the final version was published as the 14th revision of the Model Regulations [1]. These regulations included major changes that were proposed by the World Organization for Animal Health (OIE) and others to facilitate movement of diagnostic specimens from animals. The International Civil Aviation Organization (ICAO) develops the technical instructions for the safe transport of dangerous goods by air based on the Model Regulations. The International Air Transport Association (IATA) dangerous goods regulations, which are published each year, are the interpretation of ICAO instructions applied to shipments by air. The January 2006 IATA publication included the components of 14th revision of the UN SCETDG Model Regulations [2]. These regulations, as they apply to shipping diagnostic specimens, have been incorporated into the OIE *Manual of Diagnostic Tests and Vaccines for Terrestrial Animals* [3]. In addition, detailed guidelines for shipping infectious substances have been published by the World Health Organization [4]. The UN SCETDG is considering revisions of the 14th Model Regulations, which will be incorporated in the 2008 IATA regulations; however, it appears the changes will be relatively minor. These IATA regulations combined with OIE and National requirements help provide biosecurity protection. The World Health Organization (WHO) has also developed biosecurity guidelines which include biosecurity related to movement of dangerous goods [5]. The shipping regulations provide a method to help insure accountability of valuable biological material. In addition, the permit requirements for importation of infectious substances prevent introduction of foreign agents without permission.

Regulations for Shipment of Biological Material

The 14th Model Regulations and the 2006 IATA Regulations form the standard for shipping diagnostic specimens and other potentially infectious agents [1,2]. These Regulations have also been outlined in World Health Organization and OIE publications [3,4]. They serve as basis for biocontainment of this material during shipment and assist with insuring the biosecurity of infectious agents. These regulations address the shipment of infectious substances, which are defined as: *"Substances which are known or are reasonably expected to contain pathogens."* Pathogens are defined as: *"micro-organisms (including bacteria, viruses, rickettsiae, parasites, fungi) and other agents such as prions, which can cause disease in humans or animals."* Infectious substances are divided into Category A and B. Category A is defined as: *"An infectious substance which is transported in a form that, when exposure to it occurs, is capable of causing permanent disability, life-threatening or fatal disease in otherwise healthy humans or animal. (UN 2814 human or zoonotic, 2900 animal)." Category B is defined as: "An infectious substance which does not meet the criteria for inclusion in Category A (UN3373)."*

Category A Shipments

The pathogens listed in Tables 1 and 2 are designated as Category A. In addition, all new and emerging pathogens are also to be designated as Category A. As outlined in Tables 1 and 2, shipments containing cultures of these agents must be designated as Category A. Cultures are defined as follows: "*Cultures are the result of a process by which pathogens are intentionally propagated. This definition does not include patient specimens.*" The only zoonotic or animal pathogens for which both cultures and tissue samples must be included as Category A are Crimean-Congo hemorrhagic fever virus, monkeypox virus, and Nipah virus. Shipment of any material that might contain one of these pathogens, including diagnostic samples, must be designated Category A. Shipment of tissue samples for the diagnosis of all other listed zoonotic and animal pathogens are designated as Category B.

Table 1: Indicative examples of human pathogens forbidden as diagnostic specimens that must be shipped as infectious substances affecting humans (UN 2814)

Bacillus anthracis (cultures only)	Japanese encephalitis virus (cultures only)
Brucella abortus (cultures only)	Junin virus
Brucella melitensis (cultures only)	Kyasanur Forest disease virus
Brucella suis (cultures only)	Lassa virus
Burkholderia mallei – Pseudomonas mallei – Glanders (cultures only)	Machupo virus
Burkholderia pseudomallei – Pseudomonas pseudomallei (cultures only)	Marburg virus
Chlamydia psittaci – avian strains (cultures only)	*Mycobacterium tuberculosis* (cultures only)
Clostridium botulinum (cultures only)	Monkeypox virus
Coccidioides immitis (cultures only)	Nipah virus
Coxiella burnetii (cultures only)	Omsk hemorrhagic fever virus
Crimean-Congo hemorrhagic fever virus	Poliovirus (cultures only)
Dengue virus (cultures only)	Rabies virus (cultures only)
Eastern equine encephalitis virus (cultures only)	*Rickettsia prowazekii* (cultures only)
Escherichia coli, verotoxigenic (cultures only)	*Rickettsia rickettsii* (cultures only)
Ebola virus	Rift Valley fever virus (cultures only)
Flexal virus	Russian spring-summer encephalitis virus (cultures only)
Francisella tularensis (cultures only)	*Sabia* virus
Guanarito virus	*Shigella dysenteriae* type 1 (cultures only)
Hantaan virus	Tick-borne encephalitis virus (cultures only)
Hantavirus causing haemorrhagic fever with renal syndrome	Variola virus
Hendra virus	Venezuelan equine encephalitis virus (cultures only)
Hepatitis B virus (cultures only)	West Nile virus (cultures only)
Herpes B virus (cultures only)	Yellow fever virus (cultures only)
Human immunodeficiency virus (cultures only)	*Yersinia pestis* (cultures only)
Highly pathogenic avian influenza virus (cultures only)	

Table 2: Indicative examples of animal pathogens forbidden as diagnostic specimens that must be shipped as infectious substances affecting animals (UN 2900)

African swine fever virus (cultures only)	Peste des petits ruminants virus (cultures only)
Avian paramyxovirus Type 1 – Velogenic Newcastle disease virus (cultures only)	Rinderpest virus (cultures only)
Classical swine fever virus (cultures only)	Sheep-pox virus (cultures only)
Foot and mouth disease virus (cultures only)	Goatpox virus (cultures only)
Lumpy skin disease virus (cultures only)	Swine vesicular disease virus (cultures only)
Mycoplasma mycoides – Contagious bovine pleuropneumonia (cultures only)	Vesicular stomatitis virus (cultures only)

A Shipper's Declaration of Dangerous Goods must be completed for all Category A shipments and the shipper must have received training on IATA shipping regulations. The package must be labelled as "Infectious Substance Affecting Humans (UN 2814) or Infectious Substance Affecting Animals (UN 2900) and packed as outlined in the IATA regulations (Fig.1). The reader is referred to the UN Model and IATA regulations for the complete details of the shipping requirements [1,2].

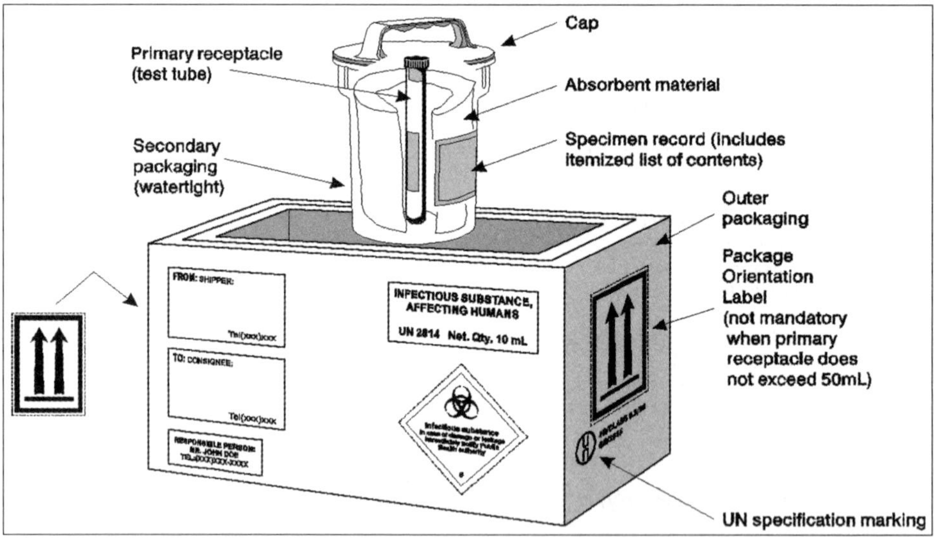

Fig. 1: Example of triple packaging system (UN P620) for the packaging and labelling of Category A infectious substances (Figure kindly provided by IATA, Montreal, Canada)

The UN Model Regulations apply to shipments other than by air [1].

Category B Shipments

As stated previously, those substances that do not meet the Category A requirement can be shipped as Category B (UN 3373). This designation can be used for shipment of all tissue specimens for diagnostic purposes from suspected infectious diseases of animals other than the three zoonotic diseases. Cultures of organisms that do not fit into the definition of Category A infectious substance can be transported as Category B. A Declaration of Dangerous Goods and training on the IATA regulations are not required. The packaging requirements (P650) are slightly less restrictive than for Category A (Fig.2). Category B shipments shall be labelled *"BIOLOGICAL SUBSTANCE, CATEGORY B"* and contain the UN3373 label. The reader is referred to the UN Model and IATA regulations for the complete details of the shipping requirements [1,2].

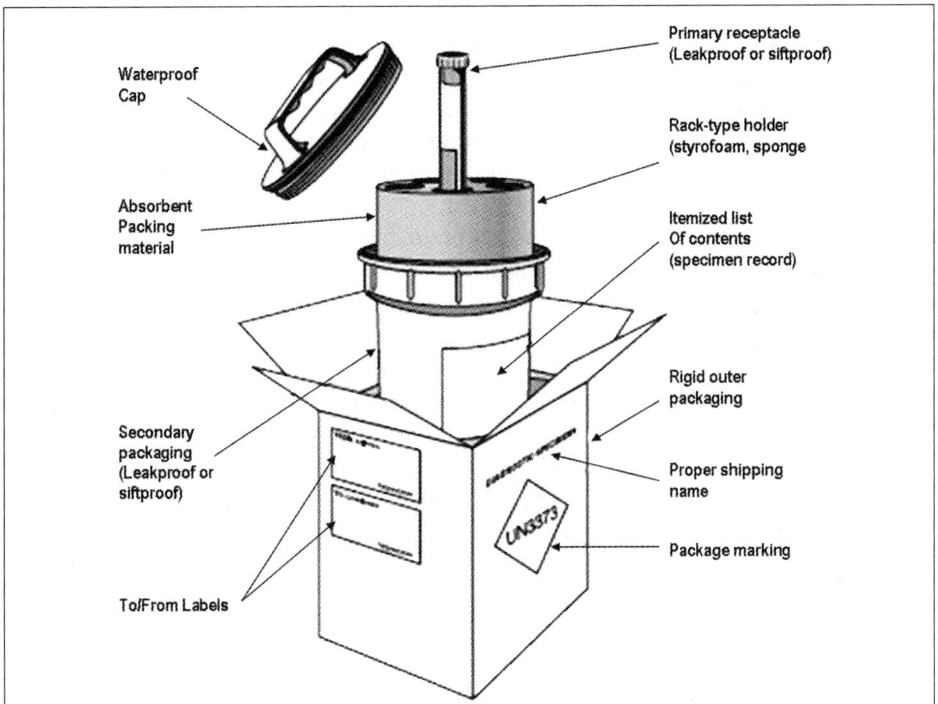

Fig.2: Example of the triple packaging system (UN650) for the packing and labelling of Category B infectious substances (Figure kindly provided by IATA, Montreal, Canada)

Packaging of Category B Shipments

As most of the diagnostic samples are shipped as Category B shipments, the following is an outline of these P650 shipping requirements (Fig.2).

i) Infectious substances assigned to UN 3373 "BIOLOGICAL SUBSTANCE, CATEGORY B" must be packed in good quality packaging, which must be strong enough to withstand the shocks and loadings normally encountered during transport. Packaging must be constructed and closed so as to prevent any loss of contents which might be caused under normal conditions of transport.

ii) The packaging must consist of three components:
- a primary receptacle;
- a secondary packaging;
- a rigid outer packaging.

iii) For liquid substances:
- the primary receptacle(s) must be leak-proof and must not contain more than 1 litre; the secondary packaging must also be leak-proof;
- adequate adsorbent material must be packed around the primary receptacle(s) to absorb all the fluid in the primary receptacle(s);
- if multiple primary receptacles are used they should be individually wrapped or separated so as to prevent contact;
- the primary receptacle or the secondary packaging must be capable of withstanding without leakage an internal pressure of 95 kPa in the range of –40°C to 55°C (–40°F to 130°F);
- the outer packaging must not contain more than 4 litres. This quantity excludes ice, dry ice, or liquid nitrogen when used to keep specimens cold.

iv) For solid substances:
- the primary receptacle(s) must be sift-proof and must not exceed the outer packaging weight limit; the secondary packaging must be sift-proof;
- adequate adsorbent material must be packed around the primary receptacle(s) to absorb all the fluid in the primary receptacle(s);
- except for packages containing body parts, organs or whole bodies, the outer packaging must not contain more than 4 kg. This quantity excludes ice, dry ice or liquid nitrogen when used to keep specimens cold;
- if there is any doubt as to whether or not residual liquid may be present in the primary receptacle during transport then packaging suitable for liquids, including absorbent materials, must be used.

v) An itemised list of contents must be enclosed between the secondary packaging and the outer packaging.

vi) If shipped at ambient temperatures or higher, the primary receptacle must have a positive means of ensuring that it is leak proof, such as a leak proof seal, heat seal or skirted stopper. If screw caps are used they should be sealed with parafilm or tape.

vii) Prefrozen packs or dry ice can be packed around the secondary receptacle. If dry ice is used, there must be an internal support to secure the secondary receptacle in the original position after the dry ice has been dissipated. The outer packaging must permit the release of carbon dioxide. There are additional requirements if liquid nitrogen is used and these are described in the Dangerous Goods Regulations.

viii) Packages containing diagnostic or clinical specimens are not required to have the net quantity marked on the outside of the package. However, where dry ice is used as a refrigerant, the net quantity of dry ice must be shown.

ix) The primary and secondary receptacles must be put into a shipping container with adequate cushioning material.

x) The packaging must be able to withstand a 1.2 metre drop test. (There are additional strength requirements for packaging used for UN 2900 and UN 2814 specimens.)

xi) At least one surface of the outer packaging must have a minimum dimension of 100 mm x 100 mm.

xii) For transport, the label 3373 must be displayed on the external surface of the outer packaging on a background of a contrasting colour and must be clearly visible and legible (Fig. 2). The mark must be in the form of a square set an angle of 45° (diamond-shaped) with each side having a length of at least 50 mm, the width of the line must be at least 2 mm, and the letters and numbers must be at least 6 mm high. The proper shipping name "BIOLOGICAL SUBSTANCE, CATEGORY B" in letters at least 6 mm high must be marked on the outer package adjacent to the diamond-shaped mark.

Exempted Substances

The UN Model Regulations and the 2006 IATA Regulations have a specific list of substances that are exempted from the Dangerous Goods Regulations. These exemptions allow the shipment of some reference reagents and samples without restriction. The following is the list of exemptions:

- *"Substances in a form that any present pathogens have been neutralised or inactivated such that they no longer pose a health risk."* The shipper is responsible for insuring that pathogens are neutralised or inactivated; if this cannot be insured the shipment will have to meet the Category A or B requirements.
- *"Substances which do not contain infectious substances or substances which are unlikely to cause disease in humans or animals"*
- *"Substances containing microorganisms which are non-pathogenic to humans or animals."*
- *"Environmental samples (including food and water samples), which are not considered to pose a significant risk of infection."*
- *"Dried blood spots, collected by applying a drop of blood on to absorbent material, or faecal occult blood screening tests and blood or blood compo-*

nents that have been collected for the purposes of transfusion or for the preparation of blood products to be used for transfusion or transplantation and any tissues or organs intended for use in transplantation."

- *"Patient specimens, for which there is minimal likelihood that pathogens are present, are not subject to these Regulations if the specimen is transported in a packaging which will prevent any leakage and which is marked with the words "EXEMPT HUMAN SPECIMEN" or "EXEMPT ANIMAL SPECIMEN" as appropriate. The packaging should meet the following conditions:*

(a) The packaging should consist of three components:
  - *(1) a leak-proof primary receptacle(s);*
  - *(2) a leak-proof secondary packaging; and*
  - *(3) an outer packaging of adequate strength for its capacity, mass and intended use, and with at least one surface having minimum dimensions of 100 mm × 100 mm;*

(b) For liquids, absorbent material in sufficient quantity to absorb the entire contents must be placed between the primary receptacle(s) and the secondary packaging so that, during transport, any release or leak of a liquid substance will not reach the outer packaging and will not compromise the integrity of the cushioning material;

(c) When multiple fragile primary receptacles are placed in a single secondary packaging, they should be either individually wrapped or separated to prevent contact between them.

Note: *In determining whether a patient specimen has a minimal likelihood that pathogens are present, an element of professional judgment is required to determine if a substance is exempt under this paragraph. That judgment should be based on the known medical history, symptoms and individual circumstances of the source, human or animal, and endemic local conditions."* This exemption is applicable to samples collected from normal animals for disease surveillance purposes.

There are also exceptions for some biological products; however, not all products are exempt and the shipper must review the regulations carefully. The following is the DGR definition of Biological Products [1,2].

"Biological products are those products derived from living organisms which are manufactured and distributed in accordance with the requirements of appropriate national authorities, which may have special licensing requirements, and are used either for prevention, treatment, or diagnosis of disease in humans or animals, or for development, experimental or investigational purposes related thereto. They include, but are not limited to, finished or unfinished products such as vaccines."

Biosecurity

Biosecurity involves the management of biological risks in a comprehensive manner to protect animal and plant life, health and protect the environment. The shipping regulations play a role in supporting biosecurity by maintaining accountability of valuable biological material (VBM). VBM may be pathogens, toxins, non-pathogenic agents, vaccine strains, genetically modified organisms, cell components and genetic elements [5]. Accountability of VBM must be maintained and the shipping regulations assist in establishing a chain of custody to support this control during transport. Biosecurity also includes prevention of or restrictions on the introduction of alien or undesirable species, which includes biological material. The OIE *Manual* outlines the requirements for import of biological material and the National requirements specify what material will be allowed [3]. If importation of potentially infectious material is allowed, it must be accompanied by an import permit.

Other Requirements and Restrictions

The following are other shipping requirements that must be complied with:

- Transport of Category A or B infectious substances in carry on luggage or transport in diplomatic pouches is strictly prohibited by international air carriers.
- Some airlines will not accept any dangerous goods (Category A or B) or require that all Category B shipments be sent as Category A. The IATA web site should be monitored to determine what airlines have additional restrictions.
- Any National requirement for importation of biological material must be met. This would include obtaining an import permit and, in some cases, an export permit. The OIE Reference Laboratories have been encouraged to establish and distribute blanket import permits in order to expedite shipments and to take other action that will expedite receiving samples.
- Cultures of the highly pathogenic agents, listed in Tables 1 and 2, must be sent as Category A. This restriction makes it more difficult and expensive to ship cultures for characterisation and epidemiological investigations. The OIE Reference Laboratories are encouraged to provide technical assistance to laboratories wishing to submit cultures for characterisation. International organisations should attempt to find financial and other assistance to insure that cultures are submitted to reference laboratories in a timely manner.

Conclusion

Regulations for the shipment of dangerous goods are developed to provide biocontainment measures that protect shipping personnel and the environment. The restrictions on Category A shipments provide the highest level of biocontainment, while Category B shipments provide a lower level, and non-infectious material is exempt. The shipping regulations assist in maintaining biosecurity for infectious substances.

References

1. United Nations Sub-Committee of Experts on the Transport of Dangerous Goods: 14th Revision of the Model Regulations for the Transport of Dangerous Goods. 2006, http://www.unece.org/trans/danger/publi/unrec/rev14/English/02E_Part2.pdf
2. International Air Transport Association: *Dangerous Goods Regulations*, 45th Edition. International Air Transport Association, 800 Place Victoria, P.O. Box 113, Montreal, Quebec H4Z 1M1, Canada, 2006; 824.
3. World Organisation for Animal Health (OIE): *Manual of Diagnostic Tests and Vaccines for Terrestrial Animals.* Chapter 1.1.1, Sampling Methods, 2005, http://www.oie.int/eng/normes/mmanual/A_00011.htm
4. World Health Organization: *Guidance on regulations for the transport of infectious substances.* 2005, http://www.who.int/csr/resources/publications/biosafety/WHO_CDS_CSR_LYO_2005_22/en/
5. World Health Organization: *Biorisk management: Laboratory biosecurity guidance.* 2006, http://www.who.int/csr/resources/publications/biosafety/WHO_CDS_EPR_2006_6/en/index.html

Dr. Jim E. Pearson, 4016 Phoenix Street, Ames, Iowa 50014, USA.
E-mail: JPearson34@aol.com

Lombard M, Dodet B, the OIE Scientific & Technical Department (eds): First International Conference of the OIE Reference Laboratories and Collaborating Centres. Dev Biol (Basel). Basel, Karger, 2007, vol 128, pp 69-72.

WAHIS and the Role of the OIE's Reference Laboratories and Collaborating Centres

K. Ben Jebara

Head, Animal Health Information Department, OIE, Paris, France

Key words: World Animal Health Information System (WAHIS), OIE Reference Laboratory, OIE Collaborating Centres, online notification system

Summary: In May 2004, OIE Member Countries adopted a resolution on the implementation of a new notification procedure. This new notification procedure led to the creation of a new World Animal Health Information System, which came into force in January 2005.

In April 2006, the OIE launched a secure web application called the World Animal Health Information System (WAHIS), an online notification system for Member Countries. For countries with fully functional internet connections, this new system has replaced the use of paper report forms. It constitutes a major achievement in the worldwide use of information technology and modern communication techniques and their applications, enabling Member Countries to notify the OIE of disease and epidemiological information in accordance with the requirements laid down in Chapter 1.1.2. of the Terrestrial Animal Health Code and Chapter 1.2.1. of the Aquatic Animal Health Code.

As well as being an online notification system, WAHIS is also a two-way communication system. It goes without saying that a good flow of information is essential if the system is to function optimally. WAHIS thus serves to disseminate information to OIE Member Countries in addition to such information being publicly disseminated via distribution lists and the World animal Health Information Database (WAHID) interface. It can also be used to ask a country to provide explanations or clarifications on the data it has submitted.

Role of the OIE's Reference Laboratories in WAHIS

There are currently 160 OIE Reference Laboratories and 20 OIE Collaborating Centres. The mandate of the OIE's Reference Laboratories, approved democratically by the Delegates of Member Countries, clearly states that Reference Laboratories may also contribute to the: "*provision of diagnostic testing facilities to Member Countries: in the case of positive results for diseases that are reportable to the OIE, the Reference Laboratory should immediately inform the OIE*

Delegate of the Member Country from which the samples originated as well as the OIE Central Bureau".

While the results of diagnostic tests performed by the OIE's Reference Laboratories should be an important source of official information, experience has shown that during the past five years the majority of the OIE's Reference Laboratories have not been complying with the above-mentioned requirement. There are many possible reasons for this. For example, they may lack a clear understanding of their mandates, they may not be fully aware of the requirement for OIE Member Countries to notify the OIE immediately of exceptional epidemiological events, or they may not wish to compromise their relationship with countries that provide them with samples.

Respect for the mandate

Table 1 lists the OIE's Reference Laboratories that complied fully with their mandates in 2005 and early 2006. During this period, only eight laboratories, representing a mere 5% of the OIE's Reference laboratories, shared the results of diagnostic tests with the OIE.

Table 1: OIE Reference Laboratories

LABORATORY	DISEASE(S)
• Veterinary Laboratories Agency (VLA), Weybridge, Surrey, UK	• Highly pathogenic avian influenza • Newcastle disease • Bovine spongiform encephalopathy (BSE)
• Friedman Shellfish Health Laboratory, School of Aquatic and Fishery Sciences, Seattle, Washington, USA	• Infection with Xenohaliotis californiensis
• Istituto Zooprofilattico Sperimentale delle Venezie, Laboratorio Virologia Padova, Italy	• Highly pathogenic avian influenza
• Institute for Animal Health, Pirbright, Surrey, UK	• Foot-and-mouth disease • Sheep pox and goat pox
• Department of Environmental and Aquatic Animal Health, Virginia Institute of Marine Science, Virginia, USA	• Infection with Haplosporidium nelsoni • Infection with Perkinsus olseni
• CSIRO, Australian Animal Health Laboratory (AAHL), Geelong, Victoria, Australia	• Highly pathogenic avian influenza
• Aquaculture Pathology Laboratory, Department of Veterinary Science and Microbiology, Tucson, Arizona, USA	• Taura syndrome
• AFSSA Sophia Antipolis, Unité Pathologie de l'abeille, Nice, France	• Bee diseases

Information on laboratories as part of the new OIE World Animal Health Information System (WAHIS)

- *Immediate notification and follow-up reports: basis of the OIE's Early Warning System*

Information on laboratories, diagnostic tests used and their results are included in the new immediate notification and follow-up reports. To ensure consistency, end users are provided with a drop-down list of diagnostic tests to choose from.

- *Annual report: Global database of national reference laboratories*

One of the changes introduced under the new World Animal Health Notification System is the compilation of data on the laboratory infrastructure in Member Countries submitted as part of the annual report procedure. Each country should provide information on its laboratory infrastructure, including the diseases in which the laboratories specialise and the tests they use, and update this information on an annual basis.

WAHIS and the Role of the OIE's Collaborating Centres

OIE Collaborating Centres are involved in horizontal activities and are not specific to one or more diseases. Centres designated in the field of epidemiology have been involved in testing the WAHIS web application during its development phase.

The OIE's Collaborating Centres have a major role to play in capacity building in the field of epidemiology and geographical information systems (GIS) for national focal points. National focal points are nominated by the Delegate of each Member Country to notify the OIE of geo-referenced epidemiological information, under the Delegate's supervision.

In view of the newly established notification procedures and the new requirements in terms of cartography and geo-referenced information, Member Countries should be prepared to improve the quality of data collected in the field. In this context, the OIE, in collaboration with its Collaborating Centre in Teramo, Italy, organised the first OIE Conference on the use of GIS in veterinary surveillance activities (Silvi Marina, Italy, 8 to 11 October 2006). The main conclusions and recommendations of this conference regarding the OIE's Collaborating Centres were as follows:

- *Capacity building*

The OIE' Collaborating Centres specialising in epidemiology, with additional geospatial expertise where necessary, should be involved in providing technical training and consultancy in the use of GIS and facilitate the sharing of experience and technical resources. Where necessary, training should also be provided to enable countries to achieve minimum animal health data standards. The Internet should be used as part of the information sharing and training activity.

- *Networking*

OIE Collaborating Centres, in close cooperation with the OIE's Central Bureau, should be used to promote and strengthen regional/international GIS-based active surveillance networks for specific diseases when needed.

Conclusion

There is a need to improve networking between the OIE and its Reference Laboratories and Collaborating Centres. The newly developed web application WAHIS offers new possibilities to achieve this.

OIE Reference Laboratories should comply more fully with their responsibility to immediately inform the OIE Delegate of the Member Country concerned as well as the OIE Central Bureau in the event of positive results of diagnostic testing of diseases in an exceptional epidemiological context. Standard operating procedures should be developed in order to build trust and satisfy the needs of all partners, while bearing in mind the importance of working towards achieving a common objective of transparency. For greater efficiency, a link should be developed between the OIE's Reference Laboratories and the WAHIS web application in order to share test results (i.e. with the country of origin and with the OIE's Central Bureau). Standardisation of the information to be provided needs to be worked out between the OIE and its Reference Laboratories.

OIE Collaborating Centres should assist in the development of training programmes and resources in the field of epidemiology and risk analysis and in the use of GIS tools by Member Countries. They should be used by the OIE to promote and strengthen regional/international active surveillance networks for specific diseases when needed, directly linked to the WAHIS web application.

Dr. Karim Ben Jebara, Head, Animal Health Information Department, OIE
E-mail: k.benjebara@oie.int

Lombard M, Dodet B, the OIE Scientific & Technical Department (eds): First International Conference of the OIE Reference Laboratories and Collaborating Centres. Dev Biol (Basel). Basel, Karger, 2007, vol 128, pp 73-78.

The New Policy of the Food and Agriculture Organization of the United Nations and its Reference Centres for the Animal Production and Health Division

J. Lubroth

Animal Health Service, Animal Production and Health Division - Food and Agriculture Organization, Rome, Italy

Key words: FAO, Reference Laboratories, Reference Centres, international collaboration, twinning.

Summary: The article explains the current procedures to be followed for institutes that are, were or would like to become Reference Centres for the Food and Agriculture Organization (FAO) of the United Nations. Within the realm of animal health many of the Reference Laboratories and Reference Centres of the World Organisation for Animal Health (OIE) are the same as those of the Animal and Health Division of FAO, particularly for diseases that are transboundary in nature, but they also address other aspects concerning health, production, standard setting, agriculture, conservation, water, and biotechnology.

Background

The Animal Health Service (AGAH) of the Animal Production and Health Division (AGA) of the Food and Agriculture Organization of the United Nations (FAO) has counted on numerous academic/research institutions, laboratories, and scientific or other similar organizations over its history working in the Organization's areas of competence to strengthen its delivery to member countries. One of the major patrons of the relationship is the group working on transboundary animal disease prevention and control. In 1994, the Director General of FAO established the EMPRES programme (Emergency Prevention System for Transboundary Animal and Plant Pest Diseases) which required additional expansion of its existing

Reference Laboratories and Collaborating Centres. Some of these institutions have been remarkable in their commitment and provision of their expertise to further animal health initiatives around the globe, and yet others have been all too silent.

Article I 2-(a) of FAO's Constitution provides that *"the Organization shall promote and, where appropriate, shall recommend national and international action with respect to: scientific, technological, social and economic research relating to nutrition, food and agriculture"* [1]. Collaboration with such institutions has been formalized through appropriate exchanges of letters or other types of agreements and instruments in line with procedures laid down in DGB 99/9 - Policy Guidelines on Preparation, Clearance and Signature of Agreements, Memoranda of Understanding and Exchange of Letters. These collaborative arrangements, whether formalized or not, have worked well in certain cases and less so in others.

From 2003 to 2006, FAO reviewed its relationship with such institutions to develop a corporate policy where the reference institution and FAO are responsible for developing collaborative plans, evaluation, and delivery. On 5 October, 2006, the FAO Director General signed the new terms for the criteria and designation for such "FAO Reference Centres".

An underlining aspect of the FAO/OIE initiative, Global Framework for the Progressive Control of Transboundary Animal Diseases, is to strengthen veterinary services to better prevent, detect and respond to incursions or interventions against epidemic animal diseases, some of which are of a transboundary nature, while others are of public health significance and yet others imply concerted campaigns to be mounted [2]. Veterinary diagnostic laboratories, sophisticated and applied research institutions, or basic and advanced epidemiological centres are critical components to make the implementation of the initiative possible.

'North-South' and 'South-South' collaborations are general terms that have been used for decades by FAO and others in the international community to reflect the partnerships between institutes in the developed world and those of underdeveloped, developing, or in-transition countries. These partnerships, when undertaken by FAO's Animal Production and Health Division, focus on specific needs and depend on its Reference Centres to mobilise technology, resources, and establish field studies or trials of mutual interest and benefit. At the OIE, these types of collaborations and assistance are termed 'twinning'.

Corporate Policy on FAO Reference Centres

FAO Reference Centres are institutions designated by the Director-General to provide specific, independent technical/scientific advice on issues related to FAO's mandate and its programme. The institution to be recommended for designation as a "FAO Reference Centre" should have demonstrated, through its previous collaboration with one or more of FAO's technical units, the following:

- active engagement in fields of expertise relevant to the work of the Organization and contribution to the implementation of its programme priorities and to strengthening capacities in countries and regions;
- ability to carry out one or several of the following functions:
  1. standardization of technology, therapeutic and other substances, and of methods/procedures;
  2. provision of reference substances and services such as quality assurance;
  3. participation in collaborative research of a scientific, technical or policy nature;
  4. contribution to capacity development (i.e., training);
  5. coordination of activities carried out by other institutions;
  6. provision of information and advice of a scientific, technical and policy nature.

In essence, therefore, a candidate FAO Reference Centre agrees to provide technical services/advice which the Organization is not equipped to provide itself.

Basic Principles and Criteria for the Selection of FAO Reference Centres

Principles

- The designation of an institution as an FAO Reference Centre will be for a specific period of time and for a defined purpose. The designation shall initially be for a maximum period of four years.
- The relevant technical unit of a Department of FAO remains responsible throughout for the management of the collaboration with the designated FAO Reference Centre.
- The designation would reflect the agreed terms of reference and the concrete areas for collaboration as laid down in a mutually agreed work plan.
- The collaboration would be subject to a joint mid-point review between the relevant technical unit and the FAO Reference Centre, with a final evaluation conducted at the end of the pre-determined period of collaboration. Upon positive review, an FAO Reference Centre could be redesignated with mutually agreed terms of reference and work plan for a specified period of time. Unless there has been any change in meeting the original criteria, this process of redesignation will be approved by the Assistant Director-General of the concerned Department.
- Institutions designated as FAO Reference Centres will, as long as this

designation remains valid, draw a number of advantages from their partnership with FAO. In particular, they will:

(i) receive international and regional acknowledgement as to their standing and high quality of work in their field of specialization;
(ii) enjoy name recognition and association with a United Nations agency, including the possibility of using FAO's name, emblem and other logos under set conditions;
(iii) have privileged access to the Organization's policy and technical information resources and benefit from FAO's experience and knowledge, at both global and country levels, in the agreed areas of collaboration;
(iv) undertake with FAO joint activities aimed at the promotion and support of future research and related training activities;
(v) conduct joint resource mobilization with FAO, where applicable; and
(vi) have access to FAO's extensive scientific and technical networks, with the possibility of broadening partnerships through collaborative efforts with other national, regional and international organizations including other FAO Reference Centres.

Criteria

Any institution recommended for designation as an FAO Reference Centre should have demonstrated, through its previous collaboration with the relevant technical unit, engagement and ability to undertake functions as described in the corporate policy document. It should also have demonstrated (i) attainment of scientific, technical and policy standing; (ii) commitment to strengthen capacity development in those areas relevant to FAO's mandate; and, (iii) prior successful collaboration with FAO for a minimum of two years (or less under special circumstances to be justified and demonstrated by the relevant technical unit) in carrying out, jointly, planned activities.

In addition, the institution should be willing to:

- use its own resources to implement collaborative activities with FAO, with the understanding that FAO is not a funding organization and is limited in financial support it can provide;
- in the interest of transparency, agree to submit a statement of Declaration of Interest, for the institution and its staff, to be completed by the head of the institution.

Procedures for Designation of FAO Reference Centres

The designation of an institution as an FAO Reference Centre shall be made by the Director-General upon the recommendation of the Assistant Director-General of the concerned Department (in the case of animal health and transboundary animal diseases, the Agriculture and Consumer Protection Department). The recommendation shall be based on a review in light of the above criteria and principles and will request the identified institution to submit a formal letter, expressing its interest in entering into a formal working relationship by being designated as an "FAO Reference Centre".

The recommendation to the Director-General should set out covering four aspects: (i) the justification and rationale for the designation; (ii) the extent and experience of prior collaboration; (iii) the scope of the envisaged collaboration; and (iv) any financial and other implications for FAO. Necessary background information will include the name, mandate and main activities of the institution, the name of its head and its source of financing.

The review would cover the legal status, competencies, operations, funding and governance arrangements of the institution. Furthermore, a prior understanding should be reached between the relevant technical unit and the institution on the areas of collaboration and functions expected to be performed in accordance with an agreed work plan. Should the institution under consideration be in any form associated directly with the private sector, it is understood that all the relevant procedures as laid down in FAO DG Bulletin 2005/28 on the Private Sector Partnerships Advisory Committee (PSPAC) would have been applied prior to the recommendation to the Director-General.

Approval in principle

When a Department of FAO would find it to be in the Organization's interest to designate a FAO Reference Centre, upon completion of the above procedures the concerned Assistant Director-General should request the approval in principle of the Director-General.

In accordance with Rule XXXVIII of the General Rules of the Organization, consultation with the Government concerned on an intended designation of a national institution as an FAO Reference Centre is required. It is proposed that such consultation takes place after the Director-General's agreement in principle is given.

Preparation, clearances and signature

Once the Director-General's approval in principle has been obtained, the relevant technical unit would prepare a draft letter of designation, in consultation with the FAO Legal Office. The draft letter of designation is then sent for clearance to the Assistant Director-General of the concerned FAO Department and Legal Office. Once all clearances are obtained, the letter of designation is submitted to the Director-General for approval and signature.

Designation

The letter of designation shall indicate the full title of the FAO Reference Centre, and clear and detailed terms of reference outlining the purpose and justification of the institutional relationship with FAO. It would: (i) give evidence of past/existing working relationships in the area(s) of mutual interest, being priority areas of FAO's programme of work; (ii) indicate in detail the area(s) of collaboration including an agreed work plan; (iii) set forth the conditions for using FAO's name, emblem and other logos following certain stipulations; (iv) clearly stipulate a process for the review of the intended institutional relationship and resulting output, with a clear timeline, indicating conditions for termination; (v) include a statement on the exclusion of any FAO legal liability for action undertaken by the FAO Reference Centre.

Disclaimer and other legal provisions

The letter of designation shall also include as an attachment a disclaimer informing, inter alia, that the FAO Reference Centre has been designated on the basis of its specific capacities and notable achievements on issues related to FAO's mandate, but that it remains a fully independent institution and that any liability which occurs during the implementation of the agreed terms of reference under the pertinent letter of designation rests with the FAO Reference Centre. It also includes specific legal provisions on inter alia intellectual property rights, FAO privileges and immunities, applicable law and settlement of disputes.

Conclusion

With the endorsement from representatives of the 198 member countries at the FAO Conference in December 2003, and the 167 delegates to the OIE General Session in May 2004, the Global Framework for the Progressive Control of Transboundary Animal Diseases is to strive to better share Reference Centres, Reference Laboratories and Collaborating Centres between the two organisations in aspects of animal health. FAO, in addition, has Reference Centres that undertake facets of animal production, processing, animal product quality and safety, which include nutritional food and other livestock goods. The selection of such Reference Centres is based on institutional depth and not dependent on the existence of an expert at a particular institution.

References

- Wu JM: Pangaea Central as the Coming Global Access to Legal, Scientific and Technical Information Through the Resource Networks of Intergovernmental Organ. INSPEL, 2001, vol 35, pp, 94-112.
- Food and Agriculture Organization of the United Nations and World Organisation for Animal Health: The Global Framework for the Progressive Control of Transboundary Animal Diseases. 2004. http://www.fao.org/ag/againfo/programmes/en/empres/documents/docs/GF-TADsMay2004.pdf

Dr. Juan Lubroth, DVM, PhD, ACVPM, Animal Health Service, Animal Production and Health Division
Food and Agriculture Organization, Rome, Italy 00153
E-mail: juan.lubroth@fao.org

Lombard M, Dodet B, the OIE Scientific & Technical Department (eds): First International Conference of the OIE Reference Laboratories and Collaborating Centres. Dev Biol (Basel). Basel, Karger, 2007, vol 128, p 79.

Session IV

New Approaches

Lombard M, Dodet B, the OIE Scientific & Technical Department (eds): First International Conference of the OIE Reference Laboratories and Collaborating Centres. Dev Biol (Basel). Basel, Karger, 2007, vol 128, pp 81-86.

Responses to a Questionnaire on Networking between OIE Reference Laboratories and OIE Collaborating Centres

G.K. Brückner, S. Linnane, F. Diaz, B. Vallat

World Organisation for Animal Health (OIE), Paris, France

Key words: Collaborating Centres, networking, OIE, reference laboratories, shipment

Summary: Two separate questionnaires were distributed to 20 OIE Collaborating Centres and 160 OIE Reference Laboratories to assess the current status of networking and collaboration among OIE Reference Laboratories and between OIE Reference Laboratories and OIE Collaborating Centres.

The questionnaire for the OIE Reference Laboratories contained 7 sections with questions on networking between laboratories, reporting of information, biosecurity quality control, and financing. Emphasis was placed in obtaining information on inter-laboratory relationships and exchange of expertise, training needs and sharing of data and information. The questionnaire for the OIE Collaborating Centres contained six sections with the emphasis on aspects related to awareness of services that can be provided, expertise that could be made available, sharing of information and the relationship with the national veterinary services of the countries concerned.

The responses to the questionnaires were collated, categorised and statistically evaluated to allow for tentative inferences on the data provided. Valuable information emanated from the data identifying the current status of networking and indicating possible shortcomings that could be addressed to improve networking.

INTRODUCTION

The World Organisation for Animal Health (OIE) has set an opportunistic goal within its 4th Strategic Plan [1] for the period 2006 to 2010 for the improvement of animal health worldwide. The mandate given to the OIE through its International Committee during the 73rd General Session in May 2005 poses a new challenge for the OIE to reinforce its capacity and for OIE Member Countries to meet these demands. An important and indispensable link in achieving the goals of the

4th Strategic Plan, is for the network of OIE Collaborating Centres and OIE Reference Laboratories to strengthen the scientific basis of the OIE for the setting of international standards for the trade in terrestrial and aquatic animals and their products and for the control of animal and aquatic diseases. It was thus essential to assess if the networking between OIE Reference Laboratories and OIE Collaborating Centres would fully support the realisation of the 4th Strategic Plan, hence the initiation of the survey to determine the status quo.

Two sets of questionnaires were distributed to 20 OIE Collaborating Centres and 160 OIE Reference Laboratories to assess the current status of networking and collaboration among OIE Reference Laboratories and between OIE Reference Laboratories and OIE Collaborating Centres. The questionnaire for the OIE Reference Laboratories contained seven sections with questions on networking between laboratories, reporting of information, biosecurity quality control, and financing. Emphasis was placed in obtaining information on inter-laboratory relationships and exchange of expertise, training needs and sharing of data and information. The questionnaire for the OIE Collaborating Centres contained six sections with the emphasis on aspects related to awareness of services that can be provided, expertise that could be made available, sharing of information and the relationship with the national veterinary services of the countries concerned. Responses to the questionnaires were received from 78.9% of the OIE Reference Laboratories and 95% of the OIE Collaborating Centres.

Responses to Selected Questions by OIE Reference Laboratories

The questions were grouped to assess networking aspects related to communication, interchange of expertise and data, logistical aspects within laboratories, and financing as experienced over the past five years. Responses were collated to reflect the combined responses to these groupings.

Networking and communication between laboratories

Visits between OIE Reference Laboratories appear not to be a common practice. A total of 79% have received less than five visits from other OIE Reference Laboratories while 83% indicated that they have not conducted more than five visits to other OIE Reference Laboratories over the past five years. Only 6% have received 10 visits or more from other OIE reference Laboratories and only 4% did more than 10 visits to other OIE Reference Laboratories.

The apparent poor contact between OIE Reference Laboratories is also reflected in the number of specimens received or submitted between them. Only 18% indicated that they have received specimens from other OIE Reference Laboratories on more than 10 occasions with 68% indicating they have received or submitted less than five specimens over the past five years and only 13% indicated the submission of specimens to other OIE Reference Laboratories on more than

10 occasions.

The exchange of data and correspondence between OIE Reference Laboratories is somewhat better with 50% indicating that they have exchanged data on less than five occasions and only 26% have corresponded on less than five occasions with other OIE Reference Laboratories in the past five years.

A total of 4% of OIE Reference Laboratories have embarked on 10 or more joint publications with other OIE Reference Laboratories while 87% produced five or less joint publications over the past five years.

The distribution of reference material between OIE Reference Laboratories appears to be more frequent with 43% indicating that they have exchanged reference material on more than 10 occasions over the past five years.

On a question asking OIE Reference Laboratories to rank their preference between the establishment of a regional network, host specific or pathogen specific networks, the majority (83%) rated a pathogen-specific network as the most desirable and of highest preference while only 20% expressed the need for a host-specific and 18% for a regional network as a high preference.

Notification of diagnostic results

Following the adoption of Resolution XXVII of the 72nd General Session of the OIE in May 2004 [2] in which it was resolved that OIE Reference Laboratories should also inform the OIE Central Bureau of results that are confirmed positive for OIE listed diseases, with the provision that such results will only be published by the OIE Central Bureau in agreement with the Delegate of the country concerned, the questionnaire aimed to assess the practical application of this Resolution. Judging from responses to the questionnaire, it appears that OIE Reference Laboratories do not readily apply the mandate given to them by this Resolution. Only 16% indicated that for specimens from their own country the OIE Central Bureau should also be notified by the laboratory and only 22% for specimens from other countries, whilst 60% indicated for specimens from there own country, that results should only be reported via the Delegate to the OIE and 59% for specimens from other countries. The most common reason given for OIE Reference Laboratories not fully implementing the Resolution is a perceived breech in the relationship of trust and confidentiality between a laboratory and its clientele.

The need was also expressed in separate comments for OIE Reference Laboratories to become more integrated and knowledgeable on the World Animal Health Information System (WAHIS). The general consensus was that this system could be used with much more benefit to OIE Reference Laboratories.

Application of biosecurity measures

A relatively high number, 65%, of laboratories indicated that biosafety and biocontainment measures and the overall management of biosecurity is either above standard or on standard while less than 5% indicated the need for improvement.

This appears to be very commendable but taking into consideration the low incidence of either reciprocal visits or inter-laboratory exchange, it could be reasoned that an evaluation of the standard of application of biosecurity measures would probably best be done by an independent evaluator such as by a parent OIE Reference Laboratory. The non-occurrence of a break in biosecurity would not necessarily be equal to the application of high or standard biosecurity measures.

A total of 70% indicated that they have employed a full-time/dedicated official to manage quality control in the laboratory.

The use of validated tests

While the concept of validated tests could be defined or perceived in different ways, it was nevertheless re-assuring that in response to the question of whether the laboratory uses validated tests for diagnostic procedures, 77% confirmed positive. The response from developing countries, however, indicated that 60% of the tests procedures used are not validated. A total of 65% indicated that they exchange reference sera with other laboratories (not necessarily OIE Reference Laboratories) for validation.

Shipment of specimens

Problems that were encountered by laboratories, especially with the shipment of specimens for confirmatory tests for Highly Pathogenic Avian Influenza, during the early phases of the H5N1 outbreaks in Asia, Europe and Africa, initiated the inclusion of a section on the shipment of specimens into the questionnaire. More than 80% of all laboratories experience serious problems with shipments by air compared to less than 10% with problems by rail and 30% encountering problems with transport by road. Whilst the OIE is already involved in negotiations to try and resolve the obstacles, it is clear that a permanent medium to long-term solution needs to be found if the rapid diagnosis of important animal diseases, especially those with a potential human/animal pathogen interface or trade sensitivity, is not to be compromised.

Financing

There appears to be an almost equal distribution between governmental and non-governmental sources of financing for laboratories. A total of 48% laboratories indicated that they are financed *in toto* from government funds while 46% indicated that they are partly funded by government and partly by non-governmental/private sources.

Responses to Selected Questions by OIE Collaborating Centres

There are several OIE Collaborating Centres that are either closely linked or on the same premises as an OIE Reference Laboratory. A close relationship and networking therefore exists in most instances of close physical or geographical proximity. In general, it appears that between OIE Collaborating Centres and OIE Reference Laboratories within countries, a reasonable level of networking has been established. However, responses to the questions indicate a need for improvement of networking between OIE Collaborating Centres and National Veterinary Services in the same country and an even greater need for networking with Veterinary Services in other countries and especially within the OIE region where the OIE Collaborating Centre is situated.

The areas of interaction between OIE Reference Laboratories and OIE Collaborating Centres listed in order of highest frequency and occurrence were: exchange of expertise (39%); exchange of information (22%); and frequent contact through reciprocal visits (16%) and collaborative projects (13%). Only 5% indicated that they have infrequent or seldom contact.

The most important areas of activity in order of priority of OIE Collaborating Centres were: development of methodologies, technical cooperation, training and harmonisation of approaches.

Whilst most of the OIE Collaborating Centres indicated that their own country is fully aware of their activities with some room for improvement in communication, the majority indicated that awareness of the role they can fulfil in both a regional and international context is poorly realised by other OIE Member Countries. This poor perception was attributed by most to lack of knowledge and awareness, but lack of funding to increase the communication strategy and outreach initiatives was also cited as an important contributing factor.

Conclusions

All respondents identified the need and wish for closer collaboration while many favoured an interchange on a pathogen-specific platform, the need to share experiences and findings, and the establishment of disease-specific networks such as for foot-and-mouth disease and bluetongue.

The questionnaire did not assess in much detail the relationship between OIE Reference Laboratories and OIE Collaborating Centres and the official Veterinary Services of OIE Member Countries. However, it would appear that there is also much need for improvement in this relationship, especially with regard to expanding the scope of involvement outside the national borders of the country in which either an OIE Reference Laboratory or OIE Collaborating Centre is situated.

As with many questionnaires of this kind, the responses need more detailed analysis and assessment on specific issues. However, judging from an overall impression, it is clear that there are definite shortcomings and a need for improvement in the communication, collaboration and sharing of expertise between OIE Reference Laboratories and between OIE Reference Laboratories and OIE Collaborating Centres. The First OIE international conference for OIE Reference Laboratories and OIE Collaborating Centres held in Florianopolis from December 3-5, 2006, is an important contributing initiative to improve the communication between these two important groups of role players in the OIE structure to provide a scientific base for almost all OIE activities and policies.

References

1. OIE, 4th Strategic Plan of the OIE: 2006 – 2010. Paris, OIE, 2005.
2. OIE, Final report of the 72nd General Session, 23 – 28 May 2004. Paris, OIE, 2004.

Dr. Gideon K. Brückner, World Organisation for Animal Health (OIE), 12 Rue de Prony 75017, Paris, France
E-mail: g.bruckner@oie.int

Lombard M, Dodet B, the OIE Scientific & Technical Department (eds): First International Conference of the OIE Reference Laboratories and Collaborating Centres. Dev Biol (Basel). Basel, Karger, 2007, vol 128, pp 87-96.

Developing Standards: the Role of the OIE Expert

T.W. Drew, D. Matthews

Veterinary Laboratories Agency (Weybridge), Addlestone, UK

Key words: OIE Reference Expert, standards, national laboratories, veterinary laboratories

Summary: The increasingly global nature of animal trade, along with greater focus on improving the production and trading opportunities for emerging nations has highlighted the need for a more standardised, harmonised approach to standardisation of diagnostics and prophylaxis. This has driven a need, on the part of both the competent authorities of member countries and international agencies, for OIE itself to develop its role as the repository and source of expertise in these fields. With this role has emerged a strong demand to develop standardisation and harmonisation of diagnostic processes, along with an ever-expanding role played by the OIE Reference Expert in networking and consultancy. This expansion comes at a price, both in terms of the burden on Reference Laboratories and the financial cost of the activities themselves. Accordingly, the role of the Reference Expert is pivotal, both in improving standards and in assuring the viability of the Reference Laboratory they represent. At a higher level, the need for strategic changes are highlighted, involving establishment of networks to share the burden and increase efficiency of delivery, along with greater interaction among international agencies and assistance in providing project-based financial support. Closer collaboration with the diagnostics and vaccine industries is also foreseen, which, if managed properly, will provide benefits to OIE Reference Laboratories through income to support their activities, to national laboratories thorough advanced, reliable and validated tests and, ultimately, to animal health itself, at the global level.

INTRODUCTION

The following thoughts on the role of the OIE expert in developing standards are based on our experiences, at the Veterinary Laboratories Agency (VLA), specifically with respect to transmissible spongiform encephalopathies (TSEs) and viruses exotic to the UK. We consider that the combination of our backgrounds in risk analysis, research and its management, and international consultancy and specialist knowledge in our subject areas of TSEs, pestiviruses and arteriviruses, provides us with a uniquely comprehensive view of the role, both for the present and the future.

This view mirrors the experiences of the other VLA experts dealing with the 11 additional diseases for which we are a named OIE Reference Laboratory and with whom we have consulted in preparing this manuscript. We therefore anticipate that there will be much agreement among the wider readership.

This introduction highlights the reality that, while the term OIE Expert covers a range of expertise, from specialist to generalist, each expert has access to additional skills amongst the scientists employed at the laboratory. The OIE expert cannot do it all on his or her own.

The Complexity of the Role

In canvassing the experts at VLA on how they interpreted the remit of the OIE Expert, and what their priorities were, we came up with a spectrum of views. These included:

"To inform and assist the world-wide scientific community, and especially that in national reference laboratories of OIE member countries, in ensuring the introduction and implementation of best practice in the detection of diseases covered by the OIE mandate";

"To provide a resource for training and act as a source of unequivocal advice and harmonised best practice in epidemiology, diagnosis and control, based on long experience at the international level";

"To provide opportunities for research collaboration and to make accrued information freely available by publication in international journals".

Named OIE Experts clearly have considerable potential to assist the OIE and the community of national reference laboratories in ensuring that high, and common, standards are achieved in the detection of diseases that impact on international trade in animals and animal products. The nature of the expert's role will, however, change with time, depending on the state of knowledge regarding individual diseases and whether dealing with an emerging disease or one that is long established with agreed methods for diagnosis and surveillance. This is reflected in the diversity of the views above.

Figure 1 attempts to capture and reflect the complexity of the role and the areas where there is an increased requirement for input in the future. These elements will now be discussed in more detail, also illustrating how each facet contributes to that role.

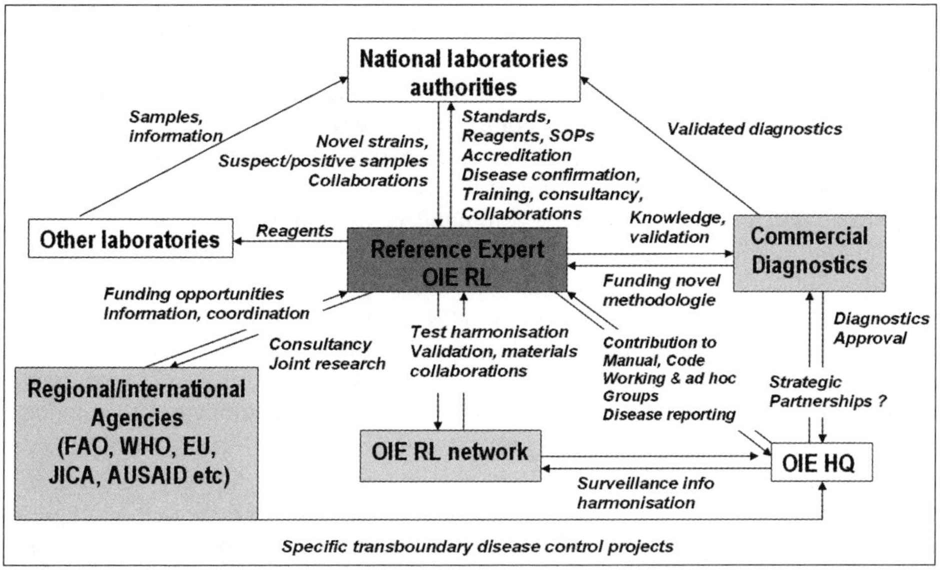

Fig. 1. Interactions of the OIE Expert. (Shaded areas represent areas of likely expansion)

THE OIE MANUAL

A key role of the OIE Expert is in the drafting or revision of relevant chapters in the *Manual of Diagnostic Tests and Vaccines for Terrestrial Animals* (or aquatic equivalent). The *Manual* has long been the traditional source of information to laboratories around the world, along with additional specific articles in OIE Scientific and Technical Reviews. Ideally such guidance should be supported by peer reviewed scientific publications of relevance to the application of diagnostic methods in the laboratory. These are usually included in the list of references in the *Manual* chapters. These publications aim to establish common diagnostic methods, and equivalence of test sensitivities and specificities around the world, although local issues may frequently intervene and compromise such attempts. The frequency of revision of the *Manual* also impacts on the currency of methodologies described – an issue the OIE has recognised and addressed by regular revisions of the online edition.

However, there are still issues to address concerning the format of chapters – some are extremely detailed, providing bench protocols for test performance, whilst others are more generic. This sometimes – but not always – reflects the state of knowledge of the pathogen and the degree to which standardised diagnosis is established.

In some fields there are signs that scientific developments may also move too rapidly to facilitate dissemination of information via the *Manual*. This is particularly so when tests are available commercially, and are evolving rapidly in response to market pressures. As a result it is neither appropriate nor possible to include details of either "the preferred test" or all test methods in the *Manual*. The consequence of such rapid development is that National reference laboratories are faced with the conflicting claims of commercial literature in making their choice of tests, and frequently require access to an impartial third party opinion on individual methods. This is certainly true of the field of TSEs at the moment, to the point where we, at the VLA, have recommended the removal of all test methods from the *Manual* during the latest revision of the chapter on the diagnosis of BSE. We believe that all the necessary information, rather than selected methods, can more readily be disseminated and accessible to all via the OIE or Reference Laboratory web sites in a form that can be updated regularly. Where such tests have been evaluated by independent bodies, links to such reports, or the original reports, can also be provided. Further assistance in interpretation can be provided by the nominated expert.

Although such provision of separate, detailed advice can be very valuable to national laboratories, it also creates the danger of non-ratified methodologies becoming established and of "independent evolution" of methodology, particularly where more than one reference laboratory exists.

The OIE Code

The capacity to conduct diagnostic tests to appropriate standards is key to supporting certification for traded products, and it is important that the rules for international trade take into account the latest knowledge on each disease, and opportunities and constraints that impact on diagnostic performance. Consequently, nominated experts should be involved, through membership of Ad Hoc groups, in the establishment and modification of trade rules in relation to specific diseases: the OIE Code. Failure to recognise scientific constraints in these key chapters will simply encourage the use of animal diseases as potential and artificial trade barriers. Ensuring the currency of the scientific basis for each chapter therefore encourages the standardisation of import/export rules for all member countries. At times however, uncertainty surrounding aspects of the pathogenesis of diseases of interest can compromise meaningful translation into trade rules. National positions may be highly selective in the presentation of data in support of change or in resisting change to the Code. While it is sometimes difficult to fully reflect the complexity of science in the Code, it is incumbent upon OIE Experts to ensure that the OIE and its supporting Commissions are fully aware of the issues, and the extent to which the Code may be dependent on interpretation of uncertain science.

Consultancy and Laboratory Services

The provision of a consultancy service for the diagnosis or confirmation of diagnosis for specific diseases is a key function of OIE Reference Laboratories.

This in itself ensures some standardisation of diagnostic quality, by limiting the number of laboratories that are involved, but the contacts that arise from the consultancy and training are potentially even more important. Ease of communication by e-mail or telephone frequently enables individuals in laboratories in different continents to compare notes, clarify issues, and resolve technical problems, thereby raising standards on a case by case basis.

Similar improvements can also arise through involvement in proficiency testing programmes, where blind panels of test samples are released to interested national or other OIE Reference Laboratories for the evaluation of test performance. Proficiency testing is often organised on a confidential basis, with results only released in a coded form, so that each laboratory can only identify itself, while comparing its performance with others involved in the distribution. Additionally, participating laboratories may wish to benefit from assistance in resolving any technical or interpretative problems that are identified during the process, possibly resulting from sub-optimal performance of their test method. Depending on the nature of the pathogen concerned there may be limitations to the extent to which positive control samples can be circulated internationally, but where it can be done the outcome can be mutually beneficial.

ISO standards are increasingly critical to guaranteeing quality of delivery and proficiency testing organised by the OIE Reference laboratories can assist in ensuring that such certification is based on more than theory. For their part, such "ring trials" can provide evidence required of such proficiency, so assisting in the attainment and retention of accreditation.

There has been some concern among Reference Experts regarding the requirement to report positive results on material from a third country directly to OIE headquarters and to the OIE Representative of that country. This causes inherent difficulties, since it can be regarded as breaching the confidentiality that exists between a testing laboratory and the commissioning body. A suitable compromise would be to initially report to the competent authority of the country concerned, also alerting them to the need to report to the OIE, with a request to acknowledge receipt, and consequent agreement, within one hour of beginning of the working day of that country, whereupon the results will be reported to the OIE. In the absence of such acknowledgement, the OIE would be notified a minimum of three hours thereafter. It would also be useful to have an "OIE Reference Laboratory Charter" which details this commitment, so that submitting laboratories are aware. Such a charter could also cover issues surrounding the ownership and downstream use of submitted material, or pathogens derived there from.

Barriers to International Harmonisation of Testing

Probably the greatest barrier to harmonisation currently existing is the difficulty in moving infectious material from one laboratory to another, across international borders. At a time when most challenges to rapid transport have been overcome, this area is becoming more legislatively challenging and, as a result, less efficient. The reasons for this are complex and are largely based on their threat as potential agro- and bioterrorist weapons and concerns over loss of intellectual property. It is imperative that the international community rapidly resolves these issues and that the OIE has the potential to play a part.

At times, politics and economics may intervene in preventing the maintenance of high standards in individual national reference laboratories. Access to the latest methodology may be blocked by conservative administrations, or lack of funding may prevent the introduction of equipment and test kits. National authorities are urged to consider the dangers of preventing their National Reference Laboratories from accessing the latest advice, and appropriate test material. Failure to detect emerging diseases can be extremely damaging, particularly so with diseases such as BSE where incursions may remain inapparent for years before rising prevalence results in recognition. Any suggestion that such incursions are missed as a result of sub-optimal testing methods can prove to be extremely damaging.

Access to appropriate reference materials is key to ensuring that test methods are optimal. Nevertheless, there are inherent dangers in relying on limited numbers of reference materials, especially if they don't fully represent the range of samples that may be encountered during the course of investigations. Strong positive or totally negative samples are inevitably necessary, but optimisation of such material can result in failure to correctly interpret equivocal cases. Consequently, it is imperative that laboratories should regularly seek to ensure that their bank of reference materials is appropriate, and to replenish stocks via OIE Reference Laboratories, and with their guidance. Where commercial test kits are now used, it is normal for control materials to be supplied with test kits, but controls may sometimes consist of recombinant proteins (e.g. TSEs) that cannot control for the entire process of sample preparation. In such circumstances, provided that testing is carried out in appropriate containment, true positive control material should occasionally be sought and used. This has not been easy with regard to the prion diseases as stocks can only be replenished by collection from naturally or experimentally infected animals, and in most countries sources are limited.

Training

Another key function of the Expert and his team is the direct training of individuals from national reference laboratories of OIE Member Countries. Experience has shown that the written word alone is frequently insufficient in ensuring appropriate delivery of diagnostic tests. Courses, held at the OIE Reference Laboratories, or occasionally at other locations more convenient for attendees, are crucial. Inevitably, the nature of the pathogen, and the level of laboratory containment and specialist equipment required when handling diagnostic material will impact on the scope for taking training to the student, as opposed to bringing the student to the source of tuition. Certain pathogens can only be handled in appropriate containment which may limit the number of locations at which training courses can be held. Many countries prohibit the importation of known positive material, presenting obstacles to training and proficiency testing. There is potential for such courses to be multidisciplinary, and to extend beyond diagnosis to epidemiology, risk assessment and control, but the key obstacle to such an approach is cost. It is normal for one or two attendees per country to participate in such courses, but they do not always have appropriate backgrounds to absorb all the skills that are to be taught.

The "Joined Up" Approach

The value of reference laboratories working as a network, rather than in competition, has long been recognised as a means to avoid duplication to achieve more than the sum of individual effort and to provide added value to the limited funding that is available. Networks need not always aim to achieve parity in all aspects of their delivery – it is possible to designate specific roles and develop the strengths within individual laboratories and dovetail the skills of the network in order to achieve a comprehensive service for a region.

The role of other international agencies in elevating competence is also increasingly important. Such agencies frequently draw on the expertise within OIE Reference Laboratories to achieve their aims. As examples, FAO, ACAIR and JICA have been making great efforts, both independently and, increasingly in partnership with each other and the OIE, in improving the diagnostic capabilities of national laboratories in diagnosis of foot-and-mouth disease and classical swine fever in Southeast Asia. The EU currently also has a project aimed at collating and sharing knowledge and experience of these two diseases. These collaborations are the way forward in regard to international disease control.

There is also an increasing awareness of the need for convergence of animal and public health. International agencies recognise this as critical to societal well-being and OIE Reference Experts are pivotal to achieving greater links with medics for those OIE-listed diseases which are also zoonotic. Such an interface will also assist us all in being anticipatory, not just reactive to emerging diseases.

Specialist Services

While the natural contact point for OIE Reference Laboratories is the National Reference Laboratory, it is not unusual for requests for guidance to be received from national authorities as well. In such instances the guidance requested may extend beyond the narrow remit of diagnostic testing, particularly into areas such as epidemiology and control of diseases. This is an important role for the OIE

Expert who can at times overcome obstacles based upon political or economic pressures that have until then prevented the National Reference Laboratory or the Competent Authority from delivering to its maximum capacity. Most importantly it is essential that advice given in such circumstances is seen to be impartial, and based solely on science with no hint of national or political influence. Experience with national rules for disease control is beneficial in offering such advice and in demonstrating how science can be translated into practice, and adds credibility to the advice provided.

Reference Laboratories can also be approached by private industry, as a source of materials or services associated with commercial test or vaccine development. This can be a dilemma for Reference Experts, who must be sensitive to the ownership issues and conditions under which material was supplied to their laboratory and to potential conflicts of interest that might arise. At VLA, such approaches are generally passed on to our commercial division, whereupon our involvement is minimised and usually restricted to acting in a consultative capacity to any study or trial that might be commissioned. A more difficult area is that associated with potential litigation associated with test or vaccine failure and we always try to limit such involvement to scientific findings, leaving interpretation or conclusion to others.

The Challenges of Standardisation

International networks and agencies have long recognised the value of partnerships in achieving the goals of the members they represent and organisations they serve. Increasingly, there is also an appreciation of their value in achieving standardisation of testing.

Indeed, a recent Canadian FIA/USDA Foresight Project report (unpublished) has stated that "*Failure to facilitate standardisation will inevitably be disastrous in the context of increasing globalisation of trade*". Such standardisation is a continuous process, requiring a committed partnership, not only between national laboratories and OIE Reference Laboratories, but also among the OIE Reference Laboratories themselves, the principle behind the OIE Network initiative. This standardisation needs continual refinement, since there is no such thing as stasis in biology. Pathogens are in constant change, driven by their evolution and by the changing environment in which they find themselves; the change is sometimes opportunistic and sometimes a response to our own attempts at control. Additionally, novel technologies come to the fore, some of which can provide certain advantage but which also require constant monitoring if their fitness for purpose is to be maintained. An example is PCR, which, although exquisitely sensitive, is vulnerable to the emergence of diverse strains which are not recognised by current primer sets.

The Issue of Equivalence

It is also worth touching on the issue of equivalence of OIE Reference Laboratories themselves. To the uninitiated it might be expected that approval as an OIE Reference Laboratory signifies equivalence. If they are to act as leaders in raising standards in National Reference Laboratories, it is appropriate that OIE Reference Laboratories should be seen to perform to an equivalent standard, at the cutting edge of science. In reality there appear to be no mechanisms by which quality of performance is judged. This fails to acknowledge the fact that skills can be lost with time, just as they must be acquired in the face of disease. Retirement of the OIE Expert is acknowledged as a factor that may influence delivery of expertise, however, the expertise supplied is not solely provided by the named expert. Logic suggests that at the very least, if OIE Reference Laboratories are to offer proficiency testing service and consultancy to National Reference Laboratories, they should themselves be subject to benchmarking by equivalent means. Inevitably such procedures take time and can be expensive, but in such circumstances it is important that all receiving countries, whether hosts to OIE or National Reference Laboratories, recognise the benefits of accessing expertise elsewhere, rather than having to invest at home. In such circumstances, chargeable proficiency schemes represent a least cost approach to maintaining national testing capacity, and may be critical to retention of OIE Reference Laboratory expertise.

The Challenge of Funding

The subject of funding is becoming increasingly important. There is little, if any direct funding from the OIE itself, so OIE Reference Laboratories are largely dependent on support from national governments. This is becoming increasingly costly, both due to the increased demands being made on the experts and the increased infrastructural costs, incurred by the need for better containment and quality standards. This can make it difficult to retain the critical mass and skills to enable delivery beyond the national need, unless research funding is also regularly available. In this environment, it is apparent that charging for services will become increasingly necessary.

If OIE Reference Laboratories are to charge for their services (which we consider is now essential), they must be fully committed to deliver in a professional and efficient way; it can no longer be regarded as an "add-on" activity. Of course, there will always be latitude to provide small scale services as part of unfunded collaborations, but the principle of charging must be established.

The technological advances made over the last decade have also provided undreamed opportunities for laboratories with quite basic facilities to perform tests which require little in the way of technological input, since they are pre-configured and presented in kit form. Beneath them, however, lie sophisticated technologies that rely heavily on a wealth of pre-design and validation data as demonstration of their fitness-for-purpose. OIE Reference Laboratories are regularly involved in development of such tests and, given their popularity and appropriateness, it is an obvious step for the laboratories to become involved in their development and ownership. And, of course, the same can be said for vaccines. The danger here is that national laboratories might regard such moves as exploitive of material submitted, or intellectual property obtained, from analyses. It might also encourage competition among OIE Reference Laboratories, at a time when we're trying to work more closely in collaboration.

Conclusions

In conclusion, the long standing role of the OIE Expert must evolve along with scientific knowledge, technological development and the changing patterns of international trade. The demands are increasingly complex and multi-faceted, requiring a depth and breadth of knowledge that is no longer really met by being a good researcher with a good team.

The demand for improved standards in every aspect of our work now requires us to achieve a level of service that can only be met by a proactive approach to coordinated partnership, along with independent, secured funding, at least in part. We encourage and look forward to greater contact with each other, as well as participating in more cross-cutting activities with other international agencies, both in the field of animal disease control and human pathogens. In view of the reliance on national funding for the delivery of OIE Reference Laboratory expertise, flexibility is also needed in securing additional income in return for advice if sources of independent expert advice are to be sustained and enhanced.

Acknowledgments

The Veterinary Laboratories Agency gratefully acknowledges the continued support of the UK Department for Environment, Food & Rural Affairs in its role as an OIE Reference Laboratory in 15 subject areas, also supported by FAO and the EU for certain diseases. For more information, see (http://www.vla.gov.uk).

Dr. T. W. Drew, Head of Virology, Veterinary Laboratories Agency, (Weybridge), Woodham Lane, New Haw, Addlestone, Surrey, KT15 3NB, United Kingdom
E-mail: t.w.drew@vla.defra.gsi.gov.uk

Lombard M, Dodet B, the OIE Scientific & Technical Department (eds): First International Conference of the OIE Reference Laboratories and Collaborating Centres. Dev Biol (Basel). Basel, Karger, 2007, vol 128, pp 97-102.

International Reference Laboratory Networks: Challenges and Benefits

S. Edwards

President of the Biological Standards Commission, Veterinary Laboratories Agency (Weybridge), UK

Key words: Infectious diseases, international standardisation, diagnostic tests, reference laboratories, networks

Summary: A number of different models can be used to demonstrate how laboratories working together across international boundaries can generate information that cannot be achieved by any one institution working in isolation. Networks can facilitate exchanges of data, of people and of biological materials. Constraints to this process include funding, protection of intellectual property rights, communication barriers, and regulatory controls (e.g. on shipment).
Examples include informal networks set up at laboratories' own initiative, formal "Networks of Excellence" established and funded by the European Union, and global initiatives inspired by international organisations such as the OIE, FAO and WHO.

CHARACTERISTICS OF A REFERENCE LABORATORY

While many international and national authorities set formal terms of reference and mandates for reference laboratories, there are some common characteristics of an institution operating in this mode, including:

- external recognition for its competence and the services offered;
- access to scientific and veterinary expertise in the designated disease or topic;
- providing confirmatory testing for other laboratories;
- providing other specialist services such as strain typing;
- monitoring trends and patterns of disease;
- publishing reports for the benefit of the wider scientific community;
- typically being supported by an active research and development programme.

Examples of International Laboratory Networks

Informal networks

For a network to succeed it must begin with informal contacts between scientists in different institutions who have areas of common interest or concern. Recognising their shared problems and identifying joint aims they can then begin to slowly build up a network of mutual support, learning and sharing. Effective informal networks should have minimal bureaucratic intervention by institutional administrations. Each partner will sustain and provide resources for its own commitments. An example of such a network is CoVetLab (www.covetlab.org) formed by five major veterinary institutions in Europe. A key feature of such networks is that they form a good launch platform for jointly applying for funding to establish more formal networks.

COST actions

COST is an intergovernmental framework, established in 1971, for the coordination of nationally-funded research at a European level and based on a flexible institutional structure. As one of the largest frameworks for research cooperation, COST includes 34 European countries, including all 27 EU Member States. COST is an acronym for European Cooperation in the field of Scientific and Technical Research, and the individual networks are termed “Actions”. COST Actions build on informal networks (i.e. bottom-up), are flexible, are based around mutual support, learning and sharing, and provide external funding for coordination activities, meetings and workshops, including travel and accommodation costs. COST does not fund research per se. It is the experience of scientists involved in such Actions that they are highly beneficial and an excellent value for the money. Typically, they provide a platform for partners to develop collaborative research proposals.

Networks of Excellence (NoE)

This is a European Union initiative established under the Sixth Framework Programme (FP6). NoEs comprise large formal networks with a minimum of three partners, but more typically between 15 to 50 members. The concept of NoE is as an instrument for strengthening excellence by tackling the fragmentation of research, leading to a reshaping of the way research in the specific area is carried out in the future. Essential elements include the bringing together of a critical mass of resources, gathering expertise to provide leadership, training and spreading excellence beyond the network. NoEs have substantial funding (millions of Euros) which supports both research projects and integration activities. The networks are self-governing, with a formal board representing the partners. They are typically of five year duration, with the intention that on completion the network will be established and sustainable. It is too early to say whether the latter aim will be achievable. There are a number of NoEs in the veterinary field, including "MedVetNet" (zoonoses, http://www.medvetnet.org/cms/), "Epizone" (epizootic disease diagnosis and control, http://www.epizone-eu.net/default.aspx), "Neuro-Prion" (http://www.neuroprion.com/en/index.html; transmissible spongiform encephalopathies) and "VenoMyc" (http://www.ucm.es/info/venomyc/; mycobacterial diseases).

FAO/OIE collaborative networks

The two international organisations involved in animal health (FAO and the OIE) have in recent years established formal networks to bring together laboratories working on foot-and-mouth disease and on avian influenza. Again these are built upon established relationships among scientists that already know each other and work together, the aim being to provide co-ordination of effort in these high profile areas, to ensure that major tasks are fulfilled and duplication of effort avoided. Taking avian influenza as an example, the network has a name, OFFLU, and a formal structure with a Steering Committee, a Scientific Committee, a secretariat and a wider network of scientific collaborators (see http://www.offlu.net/). Further details including the opportunities presented and the constraints encountered have been presented earlier [1]. Many of the challenges mentioned below have been encountered, but perhaps the greatest barrier to effective networking has been the lack of specific funding, combined with the huge demand from many different quarters, both national and international, on the scientists involved in avian influenza work.

Eight Key Challenges for Laboratory Networks

Building a network

Of the 93 diseases for which the OIE has designated Reference Laboratories, 42 have only one designated laboratory. Although this provides a single authoritative reference point for the global community, the only way such a laboratory can establish a network is by identifying national laboratories that are active and have expertise in the specified disease. A useful model here is provided within the European Union where, for a number of important diseases, there is a designated Community Reference Laboratory (CRL), one of whose functions is to co-ordinate the network of National Reference Laboratories (NRL) from each of the Member States.

For the remaining 51 diseases, the OIE has designated more than one Reference Laboratory (in some cases up to seven). The disadvantage is that there is no natural lead laboratory. This may require the parent organisation (OIE) to nominate a lead laboratory for specific tasks. An example here is provided by FAO which has designated the Pirbright Laboratory as the "World Reference Laboratory" for foot-and-mouth disease.

Whatever the organisational structures that underpin the laboratory network, the key to success, and arguably the most important message from this paper, is that networks must be founded on personal relationships. Scientists must meet (in person), identify common interests, and build trust between individuals, not just institutions. Only when a relationship is established can communication be facilitated by additional tools such as email, or tele- or video-conferencing.

Funding

While the EU provides specific funding for the activities of its CRL, with duties clearly laid out and contracts agreed, the international organisations (FAO, the OIE, WHO) rely largely on national governments to fund their reference laboratories and collaborating centres. This gives the laboratories more freedom, but the parent organisations lack the authority to demand results, despite their high expectations. The whole system runs on goodwill, and it needs to be acknowledged that national governments are under constant funding pressure.

Priorities and commitment

Laboratories and individual scientists have many demands on their time and resources. Much of their research project work is targeted towards specific deadlines, while diagnostic and surveillance work is often demand-driven, difficult to plan and can be urgent. In contrast, reference laboratory work tends to be loosely configured and slippage is rife. This is an issue for a single laboratory, but the effect is amplified at the network level because networks only function when all the members contribute their part.

Data sharing

Countries and institutions demand that scientists protect their intellectual property (IP), because this has both political and financial implications. Furthermore, individual scientists want personal recognition for and ownership of their IP, to help their own career development and to facilitate winning of future grants. There is genuine fear of theft, plagiarism and exploitation. Social factors need to be taken into account and a balance struck between individual benefit and the wider public good. An example here is provided in the field of avian influenza where the desire of certain individuals or countries to restrict availability of strains or gene sequences has been overridden in the interests of public good as we face the potential threat of a human pandemic (see http://www.gisaid.org/).

Sharing of strains and reagents

This issue has many common threads, regarding IP, to the data sharing topic (see above). They can be addressed to some extent by the use of Material Transfer Agreements (MTA), which are now widely used and accepted in exchanging biological materials between laboratories. Internationally, there may be some difficulty mounting a legal challenge if the recipients do not respect the MTA, but this has not been a major problem.

Of greater concern are the practical considerations regarding the time and cost of preparing material and the necessity to provide quality assurance (that a sample is what it purports to be and is free from extraneous agents, for example). Above all the increasingly arduous packaging and shipment regulations for dangerous goods as determined by the United Nations are a real barrier to exchanges between laboratories. This is compounded by the need for import or holding licences from national authorities, and the bureaucratic delays by customs officials who do not understand the nature of the materials being shipped.

Provision of training

Training can be very effective when delivered through a network activity. Individual network members can deliver to their own strengths, and perhaps gain some learning themselves in other areas along the way. An agreement needs to be made on the location for training, especially the availability of appropriate scientific and domestic facilities; experts should be identified to ensure that the training is of high quality, and the training should include both theoretical and practical 'hands-on' elements.

Reference reagents

The OIE has developed guidelines, particularly for reference antisera, and identified the importance of and distinction between primary (international), secondary and working (tertiary) standards. These guidelines have been published [2] and are currently under review. There is a need to develop guidelines for other reference materials such as antigens and nucleic acids. All these materials are technically complex to prepare, especially for pathogens that have multiple strains or constant genetic and antigenic drift. Where possible such preparations should be "inactivated" but there can be difficulties achieving that without also compromising the biological activity.

All international reference materials must be fully evaluated through laboratory networks to ensure that there is consensus on their activity, and reproducibility of performance. Once again the problem of funding can impede successful evaluation. Well characterised international reference reagents, with supporting data sheets, can be endorsed by the OIE through its Biological Standards Commission, and once adopted they will be listed on the OIE website.

Standardisation of diagnostic methods

Individual laboratories inevitably favour their own methods for diagnostic testing, often having invested a considerable amount of resources and intellectual effort into their development. International standardisation is an area in which laboratory networks can come into their own by enabling comparison of different methods in different locations, ideally using a standard panel of samples. Comprehensive validation of the method(s) must be part of the process. It is important, to ensure wide acceptance, that all the key players are involved from the start in the network carrying out the evaluation.

The desired outcome of this process is an agreed method that can be formally adopted by international bodies such as the OIE, and published in their Manual [3]. Once adopted, the process does not stop and the philosophy of continual improvement needs to pervade the network and all of its participants.

Benefits of Establishing Laboratory Networks

Having successfully addressed the challenges outlined above, a successful network will demonstrate the following benefits to the wider scientific community:

- Strengthened science
  - Scientists working together
  - Personal relationships
  - Backup services between institutions
- Joint outputs (papers etc)
- Spreading excellence
  - Training
  - Communications
- Better positioning for future funding bids
- Contribution to public good
  - Public health
  - Animal health
  - Economic benefits (trade)

Acknowledgements

I am very grateful to many colleagues in the Veterinary Laboratories Agency who operate our reference laboratories and have given me the benefit of their experiences, to the staff of the Scientific & Technical Dept at the OIE, to the members of the OIE Biological Standards Commission, and to many scientific colleagues worldwide for their goodwill and generosity.

This talk was originally presented at the Conference on "Emerging Diseases: Preparedness and Implementation Issues", held in Lyon 5-8 November 2006.

See http://www.iabs.org/pdf/S2.4-Edwards.pdf

References

1. Edwards S: 'OFFLU Network on Avian Influenza.' Emerg Infect Dis 2006, 12, 1287-1288.
2. Office International des Epizooties (OIE): OIE Quality Standard and Guidelines for Veterinary Laboratories, OIE, Paris, 2002, 63 pp.
3. Office International des Epizooties (OIE): Manual of Diagnostic Tests and Vaccines for Terrestrial Animals, 5th Ed. OIE, Paris, 2004, 1178 pp.

Pr. Steven Edwards, Veterinary Laboratories Agency, New Haw, Addlestone, KT15 3NB, UK
E-mail: s.edwards@vla.defra.gsi.gov.uk

Lombard M, Dodet B, the OIE Scientific & Technical Department (eds): First International Conference of the OIE Reference Laboratories and Collaborating Centres. Dev Biol (Basel). Basel, Karger, 2007, vol 128, pp 103-112.

Experiences of an OIE Collaborating Centre in Molecular Diagnosis of Transboundary Animal Diseases: A Review

S. Belák

Joint Research and Development Division of the National Veterinary Institute and of the Swedish University of Agricultural Sciences, OIE Collaborating Centre for the Application of Polymerase Chain Reaction Methods for Diagnosis of Viral Diseases in Veterinary Medicine,Uppsala, Sweden

Key words: Molecular diagnosis, PCR, automation, validation, padlock probes, microarrays

Summary: The highly contagious transboundary animal diseases (TADs), e.g., foot-and-mouth disease (FMD), classical swine fever (CSF), African swine fever (ASF) and highly pathogenic avian influenza (HPAI) are regularly occurring and re-occurring on various continents, causing severe losses. This epidemiological situation indicates the urgent need for the development of powerful, robust and high capacity new diagnostic methods in order to detect and identify the causative agents very rapidly. This report is on the experiences of an OIE Collaborating Centre and those of the MULTIPLEX-PCR and the LAB-ON-SITE EU project consortia with the development of novel methods for the improved molecular diagnosis of a range of viral diseases. Thermal amplification based real-time PCR methods (e.g.,TaqMan, Molecular Beacons, Primer-Probe Energy Transfer, and Light Upon eXtension (LUX) fluorogenic primers), and amplification without thermocycling have been elaborated for the improved diagnosis of TADs, such as FMD, swine vesicular disease, vesicular stomatitis, CSF, ASF, HPAI and Newcastle disease (ND). The simultaneous detection of various pathogens in a disease complex is facilitated by the development of multiplex PCR packages. By introducing nucleic acid extraction and pipetting robotics, together with the multi-channel real-time PCR machines, the molecular diagnostic procedures have become rapid, robust and automated. Quality control is strengthened by special precautions to avoid false positive and false negative results. By following the steps of OIE standardisation and validation, the diagnostic PCR assays have become nationally and internationally standardised and harmonised. The development of additional methods, like padlock probes and microarrays, is further improving the arsenal of nucleic acid based novel molecular diagnostic tests for TADs.

Introduction

The contagious transboundary animal diseases (TADs), listed by the Office International des Epizooties (OIE, World Organisation for Animal Health) are regularly emerging and re-emerging all around the world. The TAD outbreaks, such as foot-and-mouth disease (FMD), classical swine fever (CSF), African swine fever (ASF), rinderpest (RP), bluetongue (BT), African horse sickness (AHS), Newcastle disease (ND) and the highly pathogenic avian influenza (HPAI) are frequently causing very high losses in many countries. This is illustrated by the fact that in such a relatively short time as the last 18 months, OIE reported the occurrences of a high number of various TAD outbreaks on several continents, e.g., FMD in Africa, Asia and South America, CSF in Africa, Asia and Europe and RP in Africa and Asia. NDV was reported to re-appear for example in Slovakia and in the Czech Republic. HPAI caused severe outbreaks in Asia, Africa and Europe, leading to the deaths of many million birds and also of mammalian species, including humans, cats and mink. The recent outbreak of Rift Valley Fever (RVF) in Kenya shows the devastating effects an uncontrolled zoonosis (animal disease transmissible to humans) can have on a country. Since the beginning of the outbreak in early December 2006, the disease has affected thousands of animals (cattle, sheep, goats and camels) and over 220 humans (World Health Organization data). For more details on emerging and re-emerging TADs, see www.oie.int.

The costs of TAD outbreaks should be viewed both in terms of efforts to bring the disease under control and the consequent loss of livelihoods. As an example, with regard to the UK foot-and-mouth disease outbreak in 2001, the cost to the public sector was estimated at over 4.5 billion euro, and the cost to the private sector at over 7.5 billion euro. The ethical problem risen under the current eradication strategy and the social consequences of the slaughter of large numbers of animals must also be taken into account when considering the effects of the diseases notifiable to the OIE: (http://www.oie.int/eng/info/hebdo/A_DSUM.htm).

Veterinary health services, research institutes and international organisations (like OIE, FAO, and IAEA) make large efforts worldwide to prevent and combat TADs. Serious efforts are focused on the early detection of the agents, considering that the timely recognition of such viral infections would prevent the spread of disease to large animal populations in huge geographic areas. Thus, the development of novel, powerful diagnostic assays is a basic issue today in veterinary research and animal healthcare. Biotechnology-based diagnosis, including molecular virology and epidemiology techniques, offers a range of new methods which can accelerate and improve the diagnosis of infectious diseases in animals and humans. The new assays provide possibilities for very rapid diagnosis, since the detection of viruses can be completed within hours or hopefully, even within minutes [1-5].

The following paper summarises new trends in the development of modern molecular techniques for the improved diagnosis of TADs. Considering that this is a vast subject, the focus here is on our own experience gained at the OIE Collaborating Centre for the Application of Polymerase Chain Reaction Methods for Diagnosis of Viral Diseases in Veterinary Medicine. The majority of the new methods were developed in collaboration with partners in the EU projects Multiplex-PCR and Lab-on-Site, see http://www.multiplex-eu.org/ and http://www.labonsite.com/. The recent summarising articles [2-5], illustrating the achievements of the partners, are further discussed and updated in the present report, which was presented at the First International Conference of OIE Reference Laboratories and Collaborating Centres, Florianopolis, Brazil, 3-5 December, 2006.

The First Generation PCR Assays

At our laboratories, the first diagnostic PCR assays were established as early as 1987, just two years after the first description of the PCR principle. In the last two decades more than 50 PCR assays have been developed and validated here. They are now in routine use in the diagnostic laboratory [2-5].

Molecular Epizootiology

The rapid phylogenetic identification and tracing of viruses is termed "molecular epizootiology." Molecular epizootiology studies were conducted, for example, when genetic variants of classical swine fever virus were identified in several countries of Central Europe and when it was hypothesised that EU and US genotypes of the porcine respiratory and reproductive syndrome virus evolved from a common ancestor found in Eastern Europe [3,5,6].

Development of Real-Time PCR Assays

The various real-time PCR assays provide a novel, powerful and rapid means for virus detection and identification [2-5,7]. The real-time PCR procedures result in greatly reduced hands-on time compared to previous PCR methods, where the products were run on agarose gels and stained for visualisation. The single-run amplification on the 96-well microtitre plate format allows for automation of the procedure. The diagnostic work can further be automated by using robotics for nucleic acid extraction and pipetting. Compared to previous amplification assays, the real-time PCR technique has a further advantage: it allows for the quantitative assessment of the targeted viral genomes [5].

Different Variants of Real-Time PCR Assays in Routine Diagnosis

Various variants of real-time PCR methods and chemistries can be chosen today, like TaqMan, Molecular Beacons (MB), scorpion primers, dual probe systems such as those utilized in the LightCycler® (Roche), dye-labelled oligonucleotide ligation (DOL), and Primer-Probe Energy Transfer System (PriProET), etc. TaqMan and MB are the most commonly used methods, but other approaches, like the PriProET system, and LUX PCR are also common today [1,5,7-9,12].

Some of these assays are able to detect as few as ten genome-copies of the viruses, which indicates very high analytical sensitivity. The various real-time PCR protocols were adapted for use in routine detection of viruses and they allow for diagnosis within approximately four hours [4,5,7-10].

At the OIE Collaborating Centre, in collaboration with partner laboratories, a wide range of various real-time PCR assays for the improved diagnosis of various TADs, such as FMD, swine vesicular disease (SVD), vesicular stomatitis (VS), CSF, ASF, BT, African horse sickness (AHS), ND and HPAI was developed (see details at http://www.multiplex-eu.org/, http://www.labonsite.com/). In comparing different approaches, TaqMan, LUX, and especially PriProET techniques were found to be robust and highly reliable for diagnostic purposes. The PriProET fluorescence resonance energy transfer detection system combines probe-based real-time monitoring of PCR amplification with confirmation of probe hybridization from the melting temperature (Tm) curve. The PriProET system allows for quantification of the specific amplicon, as the fluorescence emitted from the reporter depends directly on the amount of amplicon formed [7,8]. In contrast to PriProET and LUX, the performance of other methods, like Molecular Beacons, was rather strongly influenced by point mutations in the target sequences. It was found that even the TaqMan and Scorpion primers assays can be fragile to single point mutations, or they demand longer conserved areas for hybridisation. Since many targets, especially the RNA viruses, show a high mutation rate, we suggest first using the PriProET, the LUX PCR, and in some cases the TaqMan principle for the development of diagnostic real-time PCR assays [3-5,7,8].

A quantitative real-time PCR assay was developed for the improved diagnosis of FMD, one of the most harmful TADs. This method is based on the PriProET principle and targets the 3D gene of the virus. The assay was validated for efficacy to detect all seven known FMDV serotypes and the detection limit was below the equivalent of 10 genomic copies of the virus. Analysing recent African probang samples, the method was able to detect FMDV in materials from both cattle and buffalo. Virus detection sensitivity was similar when compared to traditional virus cultivation, but the PriProET method provides a laboratory result much faster than virus cultivation. Thus, a complex diagnosis of a new foot-and-mouth-disease outbreak can be accomplished within a few hours [7].

Real-time PCR diagnosis of the vesicular cluster: FMD, SVD, VS

Since FMDV, SVD and VS cause very similar or clinically identical vesicular lesions in swine, it is a crucial requirement to develop methods for the rapid detection and identification of these three viruses in order to assure the differential diagnosis.

For this purpose, the above mentioned PriProET diagnostic PCR was completed by a range of other real-time PCR assays, to detect all three viruses [4,5,7,8].

Multiplex PCR

The principle of multiplex PCR is to use multiple primers to allow for amplification of multiple templates within a single reaction. This is a useful and practical idea for diagnostic purposes, providing the chance to detect several infectious agents in a single assay. The "classical" gel-based PCR and the real-time PCR systems are equally suitable for designing broad-range "multiplex" PCR assays. For example, a single nasal specimen can be tested from an animal suffering from a respiratory disease, or a single rectal swab in the case of an enteritis/diarrhoea syndrome. Real-time PCR is more suitable for multiplexing than classical PCR since here the individual probes can be labelled with various fluorophores, each functioning as reporter dye for one primer-set. Since the fluorescent probes emit different colour wavelengths, real-time PCR is an excellent tool for the development of multiplex PCR assays.

Considering the advantage of this technique, various multiplex PCR assays were developed in the two EU projects for the improved diagnosis of a range of TADS, such as CSF, ASF, ND and HPAI [5,11-13].

Isothermal amplification

There are techniques which allow for the isothermal amplification of the targeted nucleic acids, for example the Invader or the Loop Mediated Isothermal Amplification (LAMP) methods. Since these techniques do not require expensive PCR machines, isothermal amplification is a powerful tool today when "front line" diagnostics are developed and the diagnosticians are coming close to the outbreaks by performing "on site" diagnosis.

As the name shows, "on site" diagnosis is one of the main goals of the LAB-ON-SITE project, where various isothermal methods are being developed, such as an Invader squared assay for the detection of ASF [14].

Portable PCR machines, towards "on site" diagnosis

As with other diagnostic techniques, there is a need to further simplify PCR assays and the ultimate aim is to use them in very basically equipped laboratories or even "on site" at the farm. Many researchers are rather sceptical about the applicability of PCR under basic laboratory or field conditions. Despite these criticisms, there are certainly great advantages to be gained from the use of portable equipment which may accelerate the diagnostic work and complement the work of the central laboratories. This is the reason that various portable PCR machines appear on the market today and very cheap thermoblock platforms are being developed for the isothermal amplification systems [5].

Robotics in nucleic acid extraction

The speed and efficiency of diagnosis is further increased by the use of various nucleic acid extraction robots [4,5].

Automated diagnosis

The simultaneous use of the robot and real-time PCR detection provides an automated diagnostic procedure. With the introduction of a pipetting robot, automation is being now further completed [5].

Controls in classical and in real-time PCR

To assure the reliability of diagnostic PCR assays, it is important to incorporate internal controls. A practical approach is when a selected fragment of the host animal genome is co-amplified as an internal control. By including such an intrinsic control with its specific reporter fluorophore, information on the sample quality and on pipetting errors is obtained. Simultaneously, the system shows the amplification of the target nucleotide sequences and provides safety for the diagnosis [2-5].

OIE rules for international standardisation and validation of PCR-based diagnostic assays

The international comparison, validation and standardisation of the diagnostic assays are basic requirements. Animal health authorities require rigorous proof that the assays used in various laboratories are as reliable as possible and give identical results. International agencies like the OIE, the joint FAO/IAEA Division, national research institutions and commercial companies make large efforts to agree on international standardisation. The OIE regularly publishes standards for the validation of diagnostic assays (http://www.oie.int/eng/publicat/en_standards.htm). Among other diagnostic assays, the validation and international standardisation of nucleic acid amplification-based diagnostic methods is a major task for these authorities. The usual practice is that the specificity and the sensitivity of the newly developed PCR assays are compared to conventional assays, like virus isolation. "In house" PCR assays will soon have to be replaced by validated and standardised procedures. The validation, standardisation and quality control of PCR-based diagnostic techniques, which are now in progress, are a major task. Our laboratory, in collaboration with other international partners, has been actively involved in the validation processes following the stages of assay validation as suggested by the OIE [15, www.oie.int].

Considering these efforts and achievements, in 2005 the OIE designated our laboratory as "OIE Collaborating Centre for the Biotechnology-based Diagnosis of Infectious Diseases in Veterinary Medicine". By acknowledging this important title, we put serious effort into the development of molecular diagnostics. Furthermore, we participate in the international standardisation and validation and provide international training programmes in molecular diagnostics to students coming from various member countries of the OIE.

Other Possibilities for Molecular Diagnosis, Present and Future Trends

The above section shortly listed the various real-time PCR assays, which provide powerful novel means for the very rapid detection and quantification of targeted viral nucleic acids in clinical specimens. It is clear today that real-time PCR assays have opened a new area of molecular diagnosis. However, an Achille's heel of the PCR-based diagnostic assays is that the detection efficiency is decreased by the high nucleotide sequence variability (mismatches) in the genomes of the various variants of the targeted viruses. A higher number of mismatches between target and primer sequences yields decreased amplification efficiency. A large target-primer discrepancy may even lead to negative PCR results. The variability of target nucleotide sequences might be associated with various phenomena, such as mutations, deletions or duplications in the genomes of the sought viruses. Similarly, weaker or negative PCR results may occur when trying to amplify the genomes of newly emerging genomic variants of a virus genus. Regarding these "vulnerable" aspects of PCR-based diagnostic assays, there is a high need to develop further approaches of molecular diagnosis. Currently, there is a strong need to further increase the number of various methods in order to strengthen the molecular diagnosis, to reduce the diagnosis time and to improve the complexity of the methods available for rapid diagnosis. In the following sections several new trends will be mentioned, including new methods developed at our laboratories.

Broadly targeted triplex real-time PCR detection of Influenza A, B and C viruses based on the amplification of genomic sequences from the nucleoprotein gene

A novel multiplex real-time PCR system was designed at our laboratories for the general and simultaneous detection of influenza A, B and C viruses. By having this broad range of detection, the assay is able to simultaneously detect many members of the *Orthomyxoviridae* family originating from animals and humans. The triplex PCR is performed in a single tube, using broadly targeted primers and probes. The "ConSort"© computer program developed by Professor J. Blomberg (Uppsala University, Sweden) indicated conserved regions in the nucleoprotein genes, based on comparative sequence analysis of several thousands of influenza virus genome sequences. These regions were used to design a triplex reverse transcription real-time PCR ("3QPCR"). It was evaluated with eight influenza A reference strains (H5N1Scotland, H5N1Turkey, H5N2, H7N1, H7N7, H9N2, H1N1 and H3N2), serial dilutions of influenza A, B & C, and synthetic DNA targets. The real-time PCR assays were able to detect 1-10 viral RNA copies per PCR assay for influenza A, B and C, when testing virus isolates and clinical specimens. The tests revealed that the 3QPCR is a reliable, sensitive, and rational method for detecting and identifying influenza viruses. Its generic nature should enable detection not only of most human influenza viruses, but also of avian and other influenza strains. This wide-range assay provides a novel detection tool for diagnostic laboratories. The broad detection range ensures that the new assay provides powerful means for the discovery of newly emerging variants of influenza viruses. This is a very important issue today, considering the complicated nature of the influenza scenario worldwide [16,17].

Novel diagnosis of influenza outbreaks by using ultra rapid PCR and nucleotide sequencing

A one-step real-time PCR assay was developed at our laboratories for the rapid and simultaneous detection of a broad spectrum of influenza viruses, including highly pathogenic avian influenza viruses. This system is based on the amplification of genomic sequences from the HA gene. A unique real-time PCR system, which uses the superconvection principle ("Superconvection QPCR"; AlphaHelix, Uppsala, Sweden) is now used both for amplification and for cycle sequencing reactions [17]. Identification of pathogenicity of avian influenza viruses (HPAI or LPAI) including sequence information on the HA gene was obtained in less than two hours. The system allows for very rapid diagnosis, based on RNA purification in 30 minutes, superconvection QPCR in 30 minutes, cycle sequencing reactions with superconvection in 20 minutes and sequencing by capillary electrophoresis (up to 700 nt) in 20 minutes. The screening of different subtypes of avian influenza in a single QPCR, followed by rapid sequencing, will allow for monitoring of the viral load of influenza strains in wild birds and in farmed poultry in a rapid and effective manner. In addition, the method could provide a rapid and highly reliable molecular diagnosis in a possible pandemic influenza A scenario by prompt detection and identification of the emerging variants of the virus.

Padlock probes and microarrays for the simultaneous molecular detection of foot-and-mouth disease, vesicular stomatitis and swine vesicular disease viruses on a single platform

Padlock probes are circularizable oligonucleotides useful for highly multiplex genetic studies. After adaptation to microarray formats or to Luminex systems, padlock probes provide a novel means of powerful and very complex novel molecular diagnosis. Such systems have the capacity to simultaneously detect thousands of different target sequences in a single multiplex array system. Each target serves as template for a padlock probe equipped with a unique tag sequence, associated to that specific target. A great advantage of the system is that the circularized probes are amplified with a single universal primer pair and the fluorescently labelled products are then sorted, using the tag sequences, on a microarray. The application of padlock probes for detection of pathogens is a very recent trend in molecular diagnosis. We developed a padlock probe package for the simultaneous and rapid detection of three viruses in the vesicular disease complex of swine, FMD, SVD and VS viruses, by using a microarray-based read-out. It was observed that reaction times were comparable to real-time PCR, but with the benefit that the presence of several viruses and their various serotypes can be analyzed within the same reaction, simultaneously [5,18]. This multiplex detection system provides a powerful novel means for the simultaneous detection of the three viruses, causing very similar symptoms [18].

Conclusion

The wide range of the novel methods listed here is certainly strengthening the arsenal of novel assays for the improved detection of TADs. By providing more robust diagnosis, by shortening the time to minutes, by simultaneously detecting a wide range of viruses or virus variants, the new assays provide a powerful novel means for diagnostic laboratories. The isothermal amplification methods, the simple thermoblocks and the portable PCR machines provide novel means for the "on site" detection of various viruses. Technically, the systems have the capacity to extend the diagnosis to a very complex form in order to simultaneously detect all pathogens in a disease complex including viruses, bacteria and parasites, and probably even to estimate the immune status of the body during infection. Following the rules of OIE validation, the new methods will be introduced worldwide, in order to combat the TADs more effectively.

Acknowledgments

Thanks are due to all my colleagues in Uppsala and to our collaborating partner groups (see http://www.multiplex-eu.org, www.labonsite.com). This work was supported by EU project grants QLK2-CT-2000-00486 MULTIPLEX PCR, SSPE-CT-2004-513645 LAB-ON-SITE, as well as by grants provided by Formas, Sida/Sarec, SLF and by SVA. The critical reading and constructive remarks of Dr. M. Hakhverdyan and Dr. I. Kiss are acknowledged.

References

1. Antal M, Farkas T, Germán P, Belák S, Kiss I: Real-time PCR detection of Newcastle disease virus using light upon extension (LUX) fluorogenic primers. J Vet Diagn Invest, 2007. In press.
2. Belák S, Thorén P: Molecular diagnosis of animal diseases. Expert Rev Mol Diagn 2001;1:434-444.
3. Belák S: The molecular diagnosis of porcine viral diseases. Acta Vet Hung 2005;53:113-124.
4. Belák S, Hakhverdyan M: Recent achievements and trends in the molecular diagnosis of bovine viral diseases--a view from the "OIE Collaborating Centre for the Application of Polymerase Chain Reaction Methods for the Diagnosis of Viral Diseases in Veterinary Medicine". Dtsch Tierärztl Wochenschr 2006;113:129-133.
5. Belák S: Molecular diagnosis of viral diseases, present trends and future aspects. A view from the OIE Collaborating Centre for the Application of Polymerase Chain Reaction Methods for Diagnosis of Viral Diseases in Veterinary Medicine. Vaccine 2006; Dec 28 [Epub ahead of print].
6. Stadejek T, Stankevicius A, Storgaard T, Oleksiewicz MB, Belák S, Drew TW, Pejsak Z: Identification of radically different variants of porcine reproductive and respiratory syndrome virus (PRRSV) in Eastern Europe: Towards a common ancestor for European and American viruses. J Gen Virol 2002;83:1861-1873.

7. Rasmussen T, Uttenthal Å, de Stricker K, Belák S, Storgaard T: Development of a novel quantitative real-time PCR assay for the simultaneous detection of all serotypes of foot-and-mouth disease virus. Arch Virol 2003;148:2005-2021.

8. Hakhverdyan M, Rasmussen T, Thorén T, Uttenthal Å, Belák S: Development of a real-time PCR assay based on Primer-Probe Energy Transfer for the detection of swine vesicular disease virus. Arch Virol 2006;151: 2365-2376.

9. Kiss I, Germán P, Sámi L, Antal M, Farkas T, Kardos G, Kecskeméti S, Dán Á, Belák S: Application of LUX (light upon extension) fluorogenic primer utilizing real-time RT-PCR for the rapid detection of avian influenza viruses. Acta Vet Hung 2006;54:525-533.

10. McKillen J, Hjertner, B, Millar A, McNeilly F, Belák S, Adair B, Allan G: Molecular beacon real-time PCR detection of swine viruses. J Virol Methods 2006 Dec 28; [Epub ahead of print]

11. Agüero M, Fernández J, Romero LJ, Zamora J Sánchez, Belák S, Arias M, Sánchez–Vizcaíno JM: A highly sensitive and specific gel-based multiplex PCR assay for the simultaneous and differential diagnosis of African swine fever and classical swine fever in clinical samples. Vet Res 2004;35:551-563.

12. Farkas T, Antal M, Sámi L, Germán P, Kecskeméti S, Kardos G, Belák S, and Kiss I: Rapid and simultaneous detection of avian influenza and Newcastle disease viruses by duplex polymerase chain reaction assay. Zoonoses Public Health 2007;54:38-43.

13. Sámi L, Ursu K, McKillen J, Kecskeméti S, Belák S, and Kiss I: Simultaneous detection of three porcine viruses by multiplex PCR. Acta Vet Hung, in press.

14. Hjertner B, Meehan B, McKillen J, McNeilly F, Belák S: Development of invader squared assay for the detection of African swine fever DNA and comparison to PCR based assays. J Vir Methods 2005;124:1-10.

15. Belák S, Thorén P: Validation and quality control of polymerase chain methods used for the diagnosis of infectious diseases. Office International des Epizooties (OIE), Manual of Diagnostic Tests and Vaccines for Terrestrial Animals (mammals, birds and bees). Fifth Edition 2004, Chapter I.1.4:30-36.

16. Mohamed N, Belák S, Czifra G, Herrmann B, Berencsi, G, Blomberg J: Broadly targeted triplex real-time PCR detection of Influenza A, B and C viruses, based on the nucleoprotein gene. Submitted for publication.

17. Belák S: Early Warning. Avian flu has spread to 51 countries—36 this year alone—many of which are densely populated and deprived. Nuclear technologies are being tapped to improve the rapid detection of avian flu and other animal diseases. IAEA Bulletin 2006;48/1:62-63.

18. Banér J, Gyarmati P, Yacoub A, Hakhverdyan M, Stenberg J, Ericson O, Nilsson M, Landegren U, Belák S: Microarray-based molecular detection of Foot-and-mouth disease, Vesicular stomatitis and Swine vesicular disease viruses, using padlock probes. J Virol Methods 2007 [Epub ahead of print].

Pr. Sándor Belák, Joint Research and Development Division of the National Veterinary Institute and of the Swedish University of Agricultural Sciences, OIE Collaborating Centre for the Application of Polymerase Chain Reaction Methods for Diagnosis of Viral Diseases in Veterinary Medicine,SE-751 89, Uppsala, Sweden
E-mail: sandor.belak@sva.se

Lombard M, Dodet B, the OIE Scientific & Technical Department (eds): First International Conference of the OIE Reference Laboratories and Collaborating Centres. Dev Biol (Basel). Basel, Karger, 2007, vol 128, p 113.

Session V

The Way Forward

Lombard M, Dodet B, the OIE Scientific & Technical Department (eds): First International Conference of the OIE Reference Laboratories and Collaborating Centres. Dev Biol (Basel). Basel, Karger, 2007, vol 128, pp 115-119.

The OIE Concept of Laboratory Twinning

E. Erlacher-Vindel, G.K. Brückner, B. Vallat

World Organisation for Animal Health (OIE), Paris, France

Key words: OIE, collaborating centres, reference laboratories, twinning

Summary: Laboratory twinning was introduced in 2002, but it was never really taken on board by either OIE Reference Laboratories or OIE Member Countries in need of twinning arrangements. Recent developments in global animal health and an increase in the potential for the rapid global spread of animal diseases and pathogens call for the development and implementation of this concept. The most effective way of detecting, diagnosing, controlling and responding to animal disease and zoonotic incursions is to assure good governance in Member Countries. For this ideal to be realised, the concept laboratory twinning was born to support the development of an even geographical distribution of expertise and Reference Laboratories to facilitate access to experts and the rapid detection and diagnosis of disease. Developing and in-transition countries are also in urgent need of better access to scientific expertise to be able to justify and formulate animal health control strategies and assist in the OIE standard setting process.

INTRODUCTION

The most effective way of detecting, diagnosing, controlling and responding to animal disease and zoonotic incursions is to assure good governance in Member Countries. Integral to good governance is the ability and capacity of all Member Countries to comply with the guidelines, recommendations and international standards of the World Organisation for Animal Health (OIE) under its mandate by the World Trade Organisation (WTO).

The OIE has therefore embarked on a unique strategic initiative to develop an assessment and evaluation system to assist countries in identifying weaknesses in their system that hinder compliance with the minimum standards, guidelines and recommendations of the OIE. This evaluation system, based on the performance, vision and strategy of a country to move towards compliance (commonly referred to as the PVS-strategy) is already in operation and has been applied successfully in a number of developing and in-transition countries [1]. The assessment system considers critical aspects of veterinary service delivery such as technical capability, human and financial capital, interaction with the private sector and the needs and problems associated with access to markets. The OIE has, by linking this initiative to its overall aim to establish awareness and acceptance of the delivery of veterinary services as an international public good, elicited major financial support for this project from the World Bank and other donors, including some Member Countries of the OIE. Integral to the assessment process is the identification of the need to establish scientific and technological expertise that will allow these countries to become self-sufficient in the early detection and diagnosis of diseases and the capacity to provide the scientific justification necessary for the certification of animals and animal products for trade.

In 2002, the International Committee of the OIE adopted a Resolution to identify the need for partnerships between OIE Reference Laboratories and Collaborating Centres, as advanced research institutes, and developing and in-transition countries [2]. The OIE also realises, however, that substance can only be given to this resolution by actively moving towards scientific capacity building with the ideal of eventually establishing focal points of expertise and more Reference Laboratories within developing countries. For this ideal to be realised, the concept of laboratory twinning was born to support the development of an even geographical distribution of expertise and Reference Laboratories in order to facilitate access to experts and the rapid detection and diagnosis of disease.

The OIE guidelines, recommendations and standards for both terrestrial and aquatic animals are based on scientific principles and thereby support the requirement of the Sanitary and Phytosanitary (SPS) Agreement of the WTO [3] that sanitary measures should be scientifically justifiable. The OIE standards are developed for application by all Member Countries; it is thus imperative that all Member Countries contribute to the development of the standards and have the capacity to debate these standards from a scientific point of view when necessary. It is therefore important that developing and in-transition countries have a veterinary scientific community capable of participating in the preparation of standards. To this end, the twinning concept will create the much needed opportunities for developing and in-transition countries to become scientifically competent to debate on equal footing on the scientific justification of standards.

Twinning: an Integral Concept to Capacity Building for the Improvement of Veterinary Services

Out of the current 168 Member Countries of the OIE, almost 72% (121) are from developing countries with different levels of scientific capacity or access to scientific expertise within their national laboratories. The majority of OIE Reference Laboratories and Collaborating Centres are situated in the remaining 47 Member Countries. All 168 Member Countries of the OIE are, however, obliged to comply or move towards compliance with the international guidelines, recommendations and standards in the OIE Codes [4] and Manuals [5] for both terrestrial and aquatic animals. The establishment of the necessary scientific expertise within these countries and even geographical distribution of scientific support for their veterinary services are critical if they are to readily apply the guidelines, recommendations and standards of the OIE.

The main objective of twinning is therefore to assist laboratories in developing or in-transition countries to build their capacity and scientific expertise with the eventual aim that some will become OIE Reference Laboratories in their own right. To practically apply this concept, a link between an existing OIE Reference Laboratory or Collaborating Centre with another laboratory in a developing or in-transition country must be established for the exchange of scientific expertise and capacity building. Taking into consideration the current geographical distribution and locations of OIE Reference Laboratories and Collaborating Centres, the twinning concept could imply a transfer of knowledge, training and expertise from the 'North' to the 'South' or from an existing OIE Reference Laboratory or Collaborating Centre of the South to another less advanced laboratory in the South applying for such assistance.

Applying the concept of laboratory twinning

The procedure to implement the concept of twinning should be flexible and adaptable to different situations, from helping to publish scientific results as a first step to achieving the level required to become an OIE Reference Laboratory. The guiding or parent OIE Reference Laboratory and the selected expert(s) from that laboratory will be the driving force that will determine the success of the agreement and they will also be responsible for the implementation and the use of financial resources under the auspices and control of the OIE.

The initiation and eventual implementation of a twinning relationship between an OIE Reference Laboratory and an applicant laboratory should preferably be done in a stepwise manner, building the capacity and expertise to reach the final phase in which a participatory laboratory becomes an OIE Reference Laboratory in its own right.

For a twinning agreement to be successful, it will be essential that:

- the Delegates of the Member Countries of the two involved laboratories and their respective directors agree to the twinning arrangement;
- the twinning arrangement addresses a scientific need relevant to the sanitary situation of the country or region of the applicant laboratory;
- there are adequate communication facilities between the laboratories and experts involved;
- the objectives, time frames and desired outcomes are realistic and attainable.

The requirements for an applicant laboratory to enter into such a twinning arrangement will be detailed by the Biological Standards Commission of the OIE, taking into consideration prevailing guidelines for applicants.

The Scientific and Technical Department of the OIE Central Bureau in close cooperation with the relevant Specialist Commissions, will manage the process by:

- informing all Delegates of Member Countries under signature of the Director General of the OIE with regard to the concept and proposed steps for laboratory twinning;
- establishing a list of existing OIE Reference Laboratory or Collaborating Centre candidates and a list of applicant laboratories interested in collaboration under the OIE twinning concept;
- collating the proposed projects and applications for twinning for submission of the dossiers to the relevant Specialist Commissions;
- initiating and facilitating negotiations between the OIE and potential financial donors to assist twinning projects;
- selecting eligible projects for financial support on the basis of the estimated budget indicated by the two collaborating partners and as accepted by the relevant OIE Commission;
- assuring that the project is implemented by the existing OIE Reference Laboratory or Collaborating Centre in close cooperation with the applicant;
- assuring that technical and financial controls outlined in the mutual agreement between the participating laboratories are applied.

Conclusion

Laboratory twinning was introduced in 2002, but it was never really taken on board by either OIE Reference Laboratories or OIE Member Countries in need of twinning arrangements. Recent developments in global animal health and an increase in the potential for the rapid global spread of animal diseases and pathogens call for the development and implementation of this concept. For the global control of animal diseases, all countries need to be in a position to rapidly detect and diagnose animal diseases and pathogens. This requires easy access to expertise and OIE Reference Laboratories. Developing and in-transition countries are in urgent need of better access to OIE support and in-country scientific expertise in order to scientifically justify and formulate animal health control strategies and to assist in the standard setting process of the OIE. The implementation of laboratory twinning could be a major contributing factor to this end.

References

1. OIE, Performance, Vision and Strategy: A tool for governance of Veterinary Services, OIE, Paris, 2007.
2. OIE, Final report of the 70th General Session, 26 – 31 May 2002, OIE, Paris, 2002.
3. WTO, WTO Agreement Series, Sanitary and Phytosanitary Measures, World Trade Organisation, Geneva, 1998.
4. OIE Terrestrial Animal Health Code, 15th Edition, OIE, Paris, 2006.
5. OIE Manual of Diagnostic Tests and Vaccines for Terrestrial Animals, 5th Edition, OIE, Paris, 2004.

Dr. Gideon K. Brückner, World Organisation for Animal Health (OIE), 12 Rue de Prony 75017, Paris, France
E-mail: g.bruckner@oie.int

Lombard M, Dodet B, the OIE Scientific & Technical Department (eds): First International Conference of the OIE Reference Laboratories and Collaborating Centres. Dev Biol (Basel). Basel, Karger, 2007, vol 128, pp 121-129.

How to Improve the Efficiency of Veterinary Laboratories

D. Cousins

OIE Reference Laboratory for Tuberculosis, Department of Agriculture and Food, South Perth, Western Australia

Key words: veterinary laboratory, management, efficiency, effectiveness, research, collaboration, alliance

Summary: Veterinary laboratories around the globe are currently facing a number of important challenges. These challenges include an ageing workforce, lack of trained veterinary specialists, increased stringency for test validation, increasing emphasis on quality management, safety standards and certified accreditation processes, increasing regulation and audit processes and high costs of replacement infrastructure. Importantly, increased collaboration and linkages are high on the agenda in most countries to enhance efficiency within the sector. In Australia, mechanisms are in place to deal with some of these issues, and it is likely similar mechanisms are being used in other countries. The issues and some possible solutions are discussed.

INTRODUCTION

Veterinary laboratories around the world are currently facing a number of important challenges, including an ageing workforce, lack of trained veterinary specialists [1,2] increased stringency for test validation, increasing emphasis on quality management, safety standards and certified accreditation processes, increasing regulation and audit processes and high costs of replacement infrastructure. In Australia, many of these issues are addressed at the national level by a laboratory network lead by laboratory Directors and key stakeholders (the Subcommittee of Animal Health Laboratory Standards – SCAHLS) [3].

Importantly, increased collaboration and linkages are high on the agenda in most countries to enhance efficiency within the sector. Nationally supported mechanisms are in place in Australia to increase public-private partnerships for research and development and education and training initiatives and for infrastructure support for research, diagnostic systems and information highways. A new strategy is being devised to increase efficiencies in veterinary diagnosis and surveillance systems. In addition, many state government animal health laboratories are moving towards increasing their critical mass and efficiencies by forming alliances with universities, private laboratories and/or public health laboratories. With modern communication technologies and a willingness to enhance cooperation and collaboration between individuals, local laboratories, and states/provinces, nationally and internationally, such models can be applied on a global scale.

This paper examines the meaning of efficiency as it relates to veterinary laboratories and considers the changing face of the business of veterinary laboratories of today. Relevant issues and challenges as well as opportunities are discussed. Australia is used as a model to discuss some initiatives that might bring positive change to the complex environment in which we work.

Efficiency Versus Effectiveness

Dictionary definitions for efficiency include 'making good, thorough or careful use of resources, not consuming extra, especially making good use of time or energy' [4] and 'production of the desired effects or results with a minimum waste of time, effort or skill' [5]. The antonym of efficiency is 'wastefulness': imprudent or excessive waste of resources.

It is surprising that these definitions fail to mention quality, since the quality of results, outputs and outcomes are imperative for veterinary laboratories. Therefore, although laboratories are more and more frequently focused on efficiency it should not be at the expense of quality. While it is important to consider efficiency it is also necessary to consider effectiveness. In fact, the three elements underpinning classic project management: time, cost and quality, combined with effective management should provide a balanced approach to improvements in efficiency and effectiveness. Cost reductions, while saving money, can have a counterproductive effect on the quality we all understand is important.

Laboratory managers know their clients' views in regard to the time it takes to get a result; they'd like it yesterday. Laboratory managers understand that when their clients consider cost, they want it to be as cheap as possible, and often when their clients consider quality, they want it to be good (but perhaps only good enough not to affect the former two, price and turn around time). Increasingly however, clients of veterinary laboratory services, and particularly government, are concerned about quality.

It is well understood that the implementation of quality systems, laboratory accreditation and proficiency testing, which are part of an international trend to improve quality, increases the cost of delivery of a laboratory service. Add to this provision of a safe workplace, sound infrastructure, recruitment and retention of experienced, motivated staff and good general work practices and costs continue to escalate. However such improvements should also lead to a corresponding improvement in quality, and quality and standardization underpin effective national and global livestock diagnosis, surveillance and international trade.

How then do we start to consider improved efficiency? First let's consider the business of veterinary laboratories, which can be varied and complex. Veterinary laboratories may be owned by government, they might be part of a university, or be owned by private interests whose real agenda is to make profit. They can have a wide customer base serving a variety of clients and they might operate in one or more areas of business. They vary in their scope depending on need and resources, and their scale may vary from local to international.

The business scope of veterinary laboratories (at least in Australia) covers a wide variety of aspects. They clearly underpin emergency animal disease (EAD) preparedness and response. Veterinary laboratories are involved in disease diagnosis, surveillance, monitoring, accreditation programs, supporting trade, supporting animal health and public health, animal production and animal welfare, as well as performing research and development in a number of areas that underpin their business or that of particular industries. Depending on the circumstances, veterinary laboratories may operate for the public good or make a profit by charging fees, and many laboratories operate in both of these sectors.

A number of elements underpin a laboratory's ability to be both efficient and effective, and in an ideal world, these include operating in an environment that is politically and economically stable. Changes of government or government policy can undermine the resources and hence stability and sustainability of veterinary laboratories. Laboratories, through their people, develop expertise and experience over periods of time, and without political stability, operating efficiencies are difficult to achieve. In an ideal world there would be sufficient resources to support people, including experienced personnel, training and professional development support, and succession planning as well as an appropriate physical infrastructure, a safe environment in which to work. In addition, to retain first-rate personnel there is a need to provide ongoing interesting, challenging and valued work. Effective and efficient laboratories will also be flexible and responsive to changing demands.

A number of major shifts are occurring in the global arena. These include the effects of globalisation and the sheer (and increasing) pace of technological change (that is occurring at a rate we can barely keep up with). There are a number of issues that are being driven at the international level, many of which revolve around developing quality systems that underpin international trade. These include determining more defined requirements for international trade, the need to standardise and manage agricultural health, the differences between nations in health status and capacity, which is in itself an obstacle to trade, the demand for excellence via quality systems and the adoption of international standards and guidelines.

Factors that Impact on Veterinary Laboratory Services

During the last decade and particularly since September 11th, a number of factors have impacted on veterinary laboratories at all levels and many of these affect both the efficiency and effectiveness of our laboratories. Today we work in an increasingly regulated environment, there is increasing emphasis not just on worker safety and bio-containment, but also on biosecurity and animal ethics and welfare issues. In some developing countries, war, famine, disease and natural disasters continue to take a toll on both the workforce and on infrastructure. In addition, many laboratories built decades ago no longer meet current regulatory requirements in terms of engineering or security, although some proactive governments are now funding strategic building programs.

In many developed countries we have an ageing workforce, and comparatively speaking, we face greater difficulties attracting young people into the science and mathematics fields. Much of this is led by 'consumerism', and the desires of young people to earn large salaries (which are certainly available in law, marketing, finance, and in some regions, the booming resource sector). In Western Australia, school leavers working in the resource sector can earn about double the income of a science graduate with a PhD and several years experience. Additionally Generation Y brings a new range of expectations and behaviours [6] to managers that cannot understand why the 'younger generation' don't want to stay in a workplace for longer than a couple of years. Many countries face a lack of veterinary specialists, especially pathologists and virologists and have to develop new strategies to both train and retain such expertise [7]. In many countries there is a declining interest in agriculture and a consequential decline in funding. Who can dispute that health, education and policing warrant additional resources? Added to this, and due to the emergence of bovine spongiform encephalopathy (BSE) and its associated human disease, foot-and-mouth disease (FMD) outbreaks in the UK and Asia, severe acute respiratory syndrome (SARS), highly pathogenic avian influenza (HPAI) and other emerging, mostly zoonotic viral diseases, and the white powder incidents, there is renewed interest in biosecurity [8]. This renewed interest presents us with a unique opportunity in time to rebuild and refocus our business.

Options for Change

A number of models for increased efficiency have been heralded and discussed, and there are examples of many of these in the global arena. Privatisation is one alternative to maintaining our veterinary laboratory services. At least one state government veterinary laboratory in Australia has been privatised and most of New Zealand's veterinary laboratories have been taken over by a private company whose business is primarily human health. Another alternative is to centralise all veterinary laboratory services. This is often held up as a way of reducing costs and duplication across a country, and has been mooted as a possibility in Australia. Integration is also being talked up in Australia; the level of integration and the benefits vary depending on who is doing the talking. In theory, it should be possible to integrate animal and public health laboratories services, although this seems an easier mission for research than diagnostic services. The 'Canadian Science Centre for Human and Animal Health' complex in Winnipeg was the first facility in the world (in 2000) to combine laboratories for human and animal disease research at the highest level of biocontainment. How well this might work in a diagnostic service remains to be seen, but there appears to be some advantages for pooling resources across these sectors for research and reference laboratories.

Some states in Australia are moving to a model where government veterinary laboratories become integrated (at least to some degree) with veterinary laboratories associated with University Veterinary Faculties. Examples of where the morphing of government and university laboratories services has been achieved in the US and Canada, with mixed success, are evident. The same is true for research in New Zealand, although a softer approach is being taken. In fact, some of these examples appear to be more about creating alliances and partnerships than full integration and there seems to be general agreement about the benefits of the alliance or cooperative model.

It is important to recognize that not every veterinary laboratory can deliver every aspect of the diversity of work required by a country; neither are there resources, or a need, to maintain the highest level of expertise in every field in each laboratory. Acceptance of this has led to the development of Centres of Excellence and Reference Laboratories. These 'specialist' Centres, often recognized as such by the expertise and reputation of certain staff, continue to play a role in specialist research and development, confirmatory diagnosis and training. Although these Centres are a good business decision nationally or globally, they are often expensive to maintain and there can be considerable debate about who will fund them. When contemplating a change to an existing model, it is important to consider the business case for the change to ensure that in striving for efficiency, we don't risk destroying the working models that have evolved over time, and end up causing market failure.

Relevant Australian Initiatives

Coordinated Research, Development and Education and Training Programs

One Australian initiative that encourages efficiency through public/private research partnerships is the Cooperative Research Centres (CRC) Programme [9]. This Programme was established in 1990 'to strengthen collaborative research links between industry, research organizations, educational institutions and relevant government agencies'. A key aim was to bring together researchers and research users, focusing on research and development and emphasising the importance of collaborative arrangements to maximise the benefits of research through an enhanced process of utilisation, commercialisation and technology transfer. It also has a strong education component with a focus on producing graduates with skills relevant to industry needs.

Since the start of the Programme, 158 Cooperative Research Centres have been established, for periods of seven years each in the first instance. There are currently 57 CRCs operating in six industry areas: manufacturing technology, information and communications technology, mining and energy, agriculture and rural based manufacturing, environment and tourism and medical science and technology. CRCs can be viewed as "virtual" organizations, typically with a small headquarters in one location and often geographically dispersed nodes with people working in the laboratories of the participants of the particular CRC; this means the funds of the CRC Programme are used for projects rather than for bricks and mortar for new buildings. By bringing researchers and research users together, CRCs are changing the way in which research is done in Australia; they are encouraging teamwork and helping to develop new linkages nationally and internationally.

Of relevance to this paper is the Australian Biosecurity Cooperative Research Centre for Emerging Infectious Disease (AB-CRC), which is currently investing AUD 70 million in increasing Australia's capacity to monitor, assess, predict and respond to emerging infectious disease threats which impact on national and regional biosecurity. The AB-CRC encompasses both veterinary and public health, aiming to increase coordination and cooperation in these areas and maximise integration of research performed in either sector. The AB-CRC funds research across three broad themes: technology to enhance detection, advanced surveillance systems, and ecology of emerging infectious disease. In addition the AB-CRC has a strong focus on education and training programs, aiming to increase the capacity and number of veterinary specialists across the biosecurity arena, including laboratory orientated specialists. The AB-CRC has a focus on adoption, and aims to maximize the impact of the research and education through its Application and Linkage Program. For further information on the AB-CRC please see the AB-CRC website [10].

Coordinated Quality Initiatives

Australia has a number of initiatives that support quality, despite the increase in costs that comes with it. These include a commitment to best practice amongst veterinary laboratories facilitated by a national committee, the Subcommittee of Animal Health Laboratory Standards (SCAHLS) [3]. SCAHLS provides national coordination and standardization for veterinary laboratories, and is the body that approves new tests for use. The primary goal of SCAHLS is to facilitate the networking of government, CSIRO, private and university animal health laboratories and to establish, implement and monitor professional and technical standards within these laboratories. The Subcommittee seeks to sustain and improve the quality of livestock and livestock products and to assure market access through the application of best practice to veterinary laboratory services. SCAHLS seeks to achieve its objectives by the adoption of world's best practice for animal health laboratories servicing the livestock industries. The primary strategy is effective communication and commitment to the mission by laboratory directors.

Laboratory accreditation is compulsory for all governmental veterinary laboratories and is facilitated by the National Association of Testing Authorities (NATA) [11]. NATA is Australia's national laboratory accreditation authority, and the largest such system in the world. Australian veterinary laboratories also have a commitment to proficiency testing schemes. The Australian National Quality Assurance Program (ANQAP) [12] was established many years prior to the establishment of compulsory laboratory accreditation and has become a cornerstone of the NATA accreditation requirements. ANQAP is now an International Veterinary Proficiency Testing Program providing services to laboratories in Australia, New Zealand, USA, China, Indonesia and Thailand.

Infrastructure, Investment and Collaboration

Australia has a number of schemes for supporting infrastructure investment and collaboration. The most recent of these is the National Collaborative Research Infrastructure Strategy (NCRIS) [13]. NCRIS is a major initiative which will provide an estimated $542 million over seven years and is aimed at providing researchers with access to the infrastructure and networks necessary to undertake world-class research. The program is funding initiatives in a wide range of areas including a 'Networked Biosecurity Framework' which covers both animal and public health sectors. Australia currently has approximately $800 million invested in animal health laboratory infrastructure and each year expends more than $120 million in laboratory operations and research. A number of initiatives are in place or being developed specifically to improve efficiency.

The National Animal Health Laboratory Strategy (NAHLS) [14] is a work in progress. A shared vision for a national animal health laboratory service has been developed. This national approach is in contrast to the way in which the national system works at the moment, with each state responsible for its own services and cooperating to deliver national requirements with no real accountability but a lot of good-will. How the national system will work in practice, what it will deliver and how it will be funded are still being negotiated, but there is hope for the addition of new money from industry to enhance national capability in an efficient and sustainable manner.

Collectively, Australia's animal health laboratories play a critical role in the maintenance and enhancement of Australia's favourable animal health status. Some States/Territories have also implemented systems to share diagnostic workloads between government and private laboratories. Some use outsourcing for particular tasks e.g. biochemistry or histology and a number are developing strategic alliances with the idea of enhancing efficiency. Reference laboratories are also used to enhance efficiency in areas requiring specialist expertise.

Conclusions

This paper has attempted to summarise some of the issues that affect efficiency in veterinary laboratories, and whether individuals work in a reference laboratory or part of a bigger business, it is anticipated that some of the discussion will have relevance to most readers. Given the current heightened importance of biosecurity, many countries are in a unique position to facilitate change. We each need to maximise the effects of any increased funding that are available to ensure we build a system that is sustainable and efficient and applicable to both local and regional (and/or international) needs.

The vision for change is that we need to build with the 3 C's: Communication, Collaboration and Coordination. We should aim to build solid formal and informal networks and this First OIE Conference for Reference Laboratories and Collaborating Centres is a step towards achieving this. We should aim to enhance both local and reference laboratory capacity; we have a funding opportunity and systems to work with. We have excellent electronic tools to work with and should be able to use them to build and enhance an effective global 'e-science community'. Virtual and network based organisations are developing and there is funding available for enhancing linkages and training. There is also the need to consider building or consolidating banks of reference materials so that when new tests are developed, there is sufficient material available to validate them, whether on a national or global level. Senior researchers and OIE experts should take a leading role in this.

The ideas presented here are not necessarily unique, and many of you, working in you own organisations will be using similar ideas to achieve sustainable and efficient outcomes. Good processes and funding mechanisms are very useful but the key to real success lies within the people who are 'working the system'. If we work together for agreed outcomes then twinning, partnering or working in alliances, any of these mechanisms that create genuine collaboration should enhance efficiency, effectiveness and sustainability for delivery of veterinary laboratory services across the globe.

Acknowledgments

The assistance of the Brazilian government for funding attendance at the First OIE Conference for Reference Laboratories and Collaborating Centres is acknowledged, along with the generosity of the Australian Biosecurity Cooperative Research Centre (to whom I am currently seconded) for allowing me to attend the Conference.

References

1. Miller R: Australian Pathogenesis – a comparison between 2001 and 2005; Proceedings of the Australian Society for Veterinary Pathology. 2006;73-74.
2. Cockerell GL, Patterson DR: Closing the supply vs demand gap for veterinary pathologists: a multifaceted problem in need of a multifaceted solution. Vet Path 2005;42:403-404.
3. SCAHLS. http://www.scahls.orglau accessed November 2006.
4. Wikipedia. http://en.wikipedia.org/wiki/Main-Page accessed November 2006.
5. Stedman's Medical Dictionary. http://www.stedmans.com/section cfm/45 accessed November 2006.
6. Armour S: Generation Y: They've arrived at work with a new attitude. USA Today, Nov 8 2005. http://www.usatoday.com/money/workplace/2005-11-06-gen-y_x.html accessed November 2006.
7. Frawley PT: Review of Rural Veterinary Services Report: Department of Education Science and Training and the Department of Agriculture, Fisheries and Forestry, Australia, 2003.
8. Scott Nolten R. http://avma.org/onlnews/javma/mar02/s030102a.asp accessed November 2006.
9. Cooperative Research Centres Programme. https://www.crc.gov.au/Information/default.aspx accessed November 2006.
10. AB-CRC. http://wwwl.abcrc.org.au/ accessed November 2006.
11. NATA. http://www.nata.asn.au/go/home accessed November 2006.
12. ANQAP. http://wwwanquap.com accessed November 2006.
13. NCRIS. http://www.ncris.dest.gov.au/ accessed November 2006.
14. NAHLS. http://www.animalhealthaustralia.com.au/aahc/programs/ahsp/lab_network.cfm accessed January 2007.

Dr. Debby Cousins, current address: Australian Biosecurity Research Centre for Emerging Infectious Diseases, Curtin University, 10 Selby St, Shenton Park, WA 6008, Australia
E-mail: debby.cousins@abcrc.org.au Fax: +61 8 9266 1650

Lombard M, Dodet B, the OIE Scientific & Technical Department (eds): First International Conference of the OIE Reference Laboratories and Collaborating Centres. Dev Biol (Basel). Basel, Karger, 2007, vol 128, pp 131-138.

Reference Laboratory Networking: The European Experience in Brucellosis

B. Garin-Bastuji

European Community & OIE/FAO Reference Laboratory for Brucellosis, French Food Safety Agency (AFSSA), Maisons-Alfort, France

Key words: Laboratory networking; European experience; Community Reference laboratory; Brucellosis

Summary: Over the past forty years, a huge effort of standardisation of tests and harmonisation of sanitary policies and requirements has been made in the European Union (EU). The EU has itself greatly evolved in its size and policies. Voluntary uncoordinated actions initially had some success but not enough to attain a level of standardisation compatible with a single market and to achieve the sanitary level foreseen by the community. The creation of a Community Reference laboratory with specific responsibilities and tasks and the provision of adequate financial assistance is expected to lead to better co-ordination of activities and to ensure international harmonisation of tests and procedures.

INTRODUCTION

Brucellosis is a zoonosis due to Gram-negative bacteria of the genus *Brucella*. It is considered a major risk to public health with an annual world incidence of 500,000 according to WHO and 1,091 human cases notified in the EU in 2003. The infection in domestic animals concerns mainly cattle, sheep, goats and pigs and is of great economical importance due to abortions, genital lesions and infertility in both sexes, milk production losses and international trade restrictions. Animal brucellosis is distributed throughout the world except for a low growing number of countries where eradication has been achieved (Fig.1). Most member states of the European Union (EU), Canada, the United States of America (USA), Japan, Australia and New Zealand are free of brucellosis in cattle, sheep, goats, and pigs while in Africa, Latin America and Asia the infection is still enzootic in at least one species. Sporadic outbreaks of porcine brucellosis, however, are regularly reported in free-range pig farms in some brucellosis free countries of Europe and North America.

Experience has shown that reference laboratories play an important role as a scientific and technical support in the area of feed and food safety and animal health. This is particularly important as problems are regularly encountered in trade of animals or products where conflicting test results may arise. Scientifically sound and uniform testing is an element of fundamental importance for appropriate disease diagnosis and for the application of the necessary control and eradication measures.

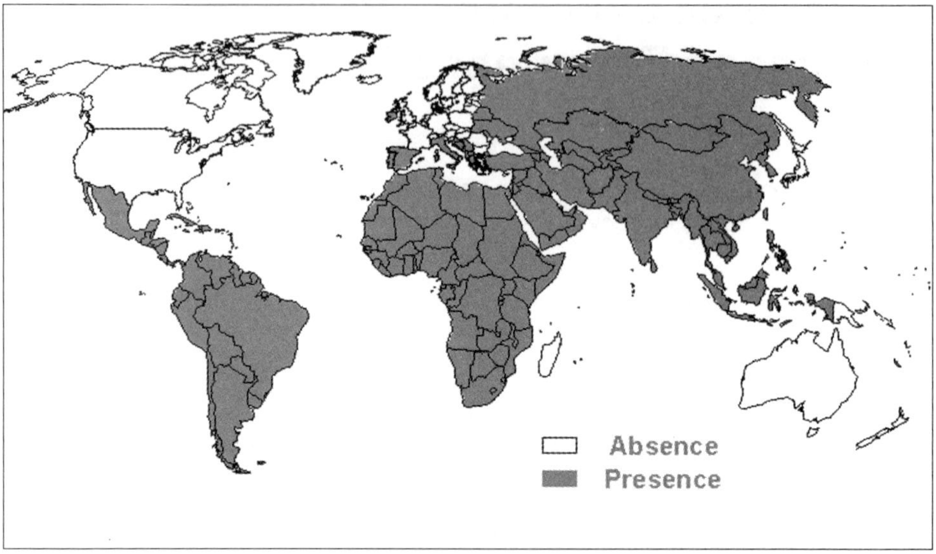

Fig. 1: World distribution of bovine, sheep and goats and/or porcine Brucellosis in 2004.

History of International Standardization in Animal Brucellosis

Serological methods used for the control and the eradication of brucellosis were among the first employed in animal health to be standardised at an international level. The first international standard serum (1st ISABS) was produced in cattle and established by WHO, FAO and the OIE in 1953 for standardizing the serum agglutination test (SAT). In 1958 WHO and FAO agreed on the conditions for standardizing the Milk ring test used for the diagnosis in bulk tank milk samples, which were defined against this standard. In 1968, a 2nd ISABS was produced in cattle by CVL (Weybridge, UK), standardised against the 1st ISABS, approved by all concerned international bodies (the OIE, FAO, WHO and the EEC) then used for standardizing the newly accepted tests, i.e. the Complement Fixation test (CFT) in 1971, the Rose-Bengal test (RBT) in 1979 and then indirect ELISA. Finally, ELISA specific bovine standard sera (strong positive and low positive) were established and conditions of use were agreed at the international level for the standardisation of all ELISA methods (i- and c-ELISA) by the OIE.

History of European Standardization in Animal Brucellosis

The European Union of today is the result of a long process of enlargement to new member states (MS) that began half a century ago. Six countries (Belgium, France, Germany, Italy, Luxembourg and The Netherlands) constituted the first union named the European Economic Community (EEC) in 1957. Fifty years later, with new adhesions in 1973 (Denmark, Eire and UK), 1981 (Greece), 1995 (Austria, Finland and Sweden), 10 new MS in 2004 and two in 2007, this union has become the European Union (EU) with a larger level of political and economical integration and a total of 27 MS. Despite - or because of - this continuous growth and evolution and due to the progressive implementation of a single free market throughout the community during these fifty years, standardisation of procedures has always been a goal in the EU. This is particularly the case in animal trade and standardization of tests for brucellosis. The 1st ISABS was adopted for SAT in the EEC in 1964 for cattle and, in 1968 the EU adopted the 2nd ISABS followed by standard protocols adopted for CFT (1971), RBT (1979) and iELISA (1991) in cattle as well as RBT and CFT in sheep and goats (1991). More recently, the European Directorate for Quality of Medicines (EDQM) established the first reference strain of the vaccine strain Rev.1 (official vaccine in small ruminants). In 2002, new standards were established for iELISA (the sole ELISA technique approved in cattle by EU for trade and surveillance) and cELISA and Brucellin skin tests were approved as complementary tests, employed in case of doubtful positive results with prescribed tests in free herds. With the exception of iELISA, the adopted standardisation was in accordance with the OIE requirements as stated in the OIE Manual.

Efforts were also made to reinforce the harmonisation of surveillance and eradication programmes. Up until the 90's, the European Commission (EC) set up formal groups of experts to help in regulations improvement. With the increasing number of MS, however, the EC has progressively implemented new tools in order to reinforce the level of harmonisation such as: reinforcement of the Food and Veterinary Office (Inspection body) in 1997, and the creation, in 2000, of a Task Force for Monitoring Eradication of Diseases with specific expert subgroups dedicated to brucellosis in cattle, sheep and goats, assisting the Commission in evaluating the respective cost-benefit ratio of MS animal disease eradication programmes co-financed by the Community. A COST action (COST Action 845) was granted from 2000 to 2006 for financially helping laboratories (applied and basic research) to meet and share their concern in the field of brucellosis in animals and humans and to improve the efficiency of control of brucellosis and its eradication in animals, thereby facilitating international trade and improving public health. Finally the creation of a Community Reference Laboratory took place in 2006. Other more basic research programmes (CAMAR 90-0018 and FAIR 95-0134, for instance) were also funded by the Commission in order to improve the standardisation of brucellosis diagnosis. All these actions have clearly helped in improving the level of harmonisation and standardisation of the sanitary policies and their specific tools (including diagnostic tests and vaccines) throughout the European Community. The main actions launched for brucellosis in ruminants are summarised in tables 1 and 2. The main action with regard to brucellosis in pigs, the importance of which is much lower in Europe - although it has increased in the past years due to the high level of infection of wild boar and hare populations - was voluntary standardisation of *Brucella suis* typing methods under the auspices of the COST action previously mentioned.

Table 1: Actions undertaken for improving harmonisation in bovine brucellosis

- 1985: CEC *B. melitensis* meeting
- EEC-CAMAR N°8001-CT90-0018 research project (91-94) "Brucellosis in small ruminants" (B, F, NL, UK) RBT, CFT, MRT, iELISA (serum & milk), BST
- Standardisation trials on RBT & CFT (F, P, Sp, UK & COST action)
- Proficiency ring-trials in the whole union (VLA) under the auspices of the EC (2000-2004)
- Collaborative ring-trial for inclusion of additional standard tests (FPA, cELISA, mRBT, AGID) (2003-2004)
- COST Action 845: OIE manual revision (2004 Ed.)
  Meta-analysis for evaluation of prescribed diagnostic tests + FPA/cELISA/AGID in sheep & goats by EFSA (2005-2006)

Table 2: Actions undertaken for improving harmonisation in sheep & goats brucellosis

- 1986 Validation ring-trials for the standardisation of RBT with ISABS (OIEISS)
- FAIR CT95–0134 research project (95-99) on "Non-specific serological reactions in bovine brucellosis" (B, F, NL, UK) iELISA (individual & pools), PCR, cELISA, BST
- Validation ring-trials (B, F, G, UK) for the standardisation of the indirect ELISA (1990-1991 & 2000-2001) with the OIE standards
- Proficiency ring-trials in the whole union (VLA) under the auspices of the EC (2000-2004).
- COST Action 845: OIE manual revision (2004 Ed.)
  Meta-analysis for evaluation of official diagnostic tests + FPA/cELISA in cattle by EFSA (2005-2006)

Present Situation of European Standardisation in Animal Brucellosis

Regulations regarding sanitary requirements for intra-community trade were established in 1964 for cattle and pigs and in 1991 for sheep and goats. These regulations have since been amended regularly. They are harmonised for standardisation and already include NRLs' responsibilities for: (i) approval of validation study results demonstrating the reliability of test methods used in the MS; (ii) determination of the maximum number of samples to be pooled in iELISA kits; (iii) calibration of the standard secondary reference national standard sera ("working standards") against OIE standard sera; (iv) quality checks of all antigens and ELISA kit batches used in the MS; and (v) co-operation within the EU Network of NRLs for Brucellosis.

However, as far as surveillance and eradication are concerned, no guidelines or rules have been established at the EU level. The situation varies considerably among MS (Fig. 2), with free zones employing surveillance programmes and more or less infected areas with on-going eradication or vaccination plans. Problems faced by the authorities are therefore highly variable depending on the MS and their needs, especially with regard to tests. While the EC is co-financing brucellosis eradication programmes, diagnostic tools that are standardised at the EU level are used for trade control, surveillance of free zones, and eradication purposes which frequently have contradictory sensitivity and specificity needs. There is great heterogeneity with regard to application of the regulations across the MS: (i) different tests and test methods/protocols are used; (ii) different protocols are used for controlling the reagents (if this is done); (iii) there is no harmonisation of national standards (if they do exist); and (iv) the EU network is not effectively managed between meetings, with no follow-up of the proficiency ring-trials.

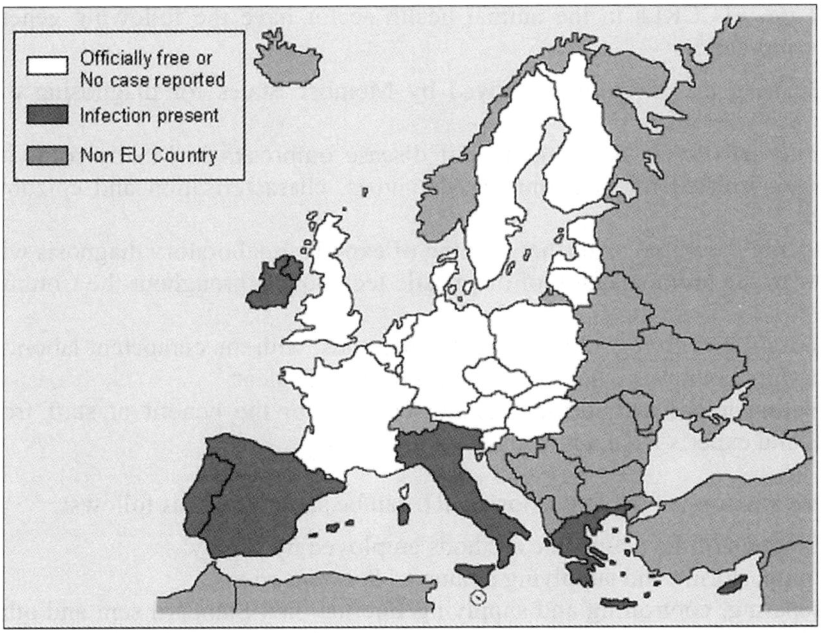

Fig. 2: Epidemiological situation of brucellosis in cattle and/or sheep and goats in EU in 2005.

In conclusion, up until this point, the work done towards harmonisation/standardisation was largely voluntary and thus not all MS were included in standardisation actions and there was no co-ordination or monitoring system in place. There was no legal chain of responsibilities (EC/MS), no specific adequate funding (except for research projects), and a consensus was not always reached. Decision-making at a Community level was difficult or almost impossible. Harmonisation became critical with an increasing number of MS and tests.

As a result, like for several other diseases and food contaminants, the creation of a Community Reference Laboratory appeared essential in order to implement a more coordinated and integrated approach together with the EC, Task Force, FVO, EDQM and EFSA. This was in line with the EC "White Paper on Food Safety" (2000) which proposed a more coordinated and integrated approach to the organisation of food safety in view of achieving the highest possible level of health protection and which recognised that the establishment of Community Reference Laboratories for new areas should be examined.

The Community Reference Laboratory (CRL) for Brucellosis

Nominated after a specific call for selection and designation with a clear definition of general functions and duties and disease-specific responsibilities and tasks, all the EU CRLs in the animal health sector have the following general functions and duties:

- coordinating the methods employed by Member States for diagnosing diseases;
- assisting actively in the diagnosis of disease outbreaks in MS by receiving pathogen isolates for confirmatory diagnosis, characterisation and epizootic studies;
- facilitating the initial or further training of experts in laboratory diagnosis with a view to the harmonisation of diagnostic techniques throughout the Community;
- collaborating, with regard to diagnostic methods, with the competent laboratories in third countries where those diseases are prevalent;
- conducting initial and further training courses for the benefit of staff from NRLs and experts from developing countries.

The future missions of the Brucellosis CRL can be summarised as follows:

(i) Coordination of the diagnostic methods employed by MS by:
  - typing, storing and supplying strains of *Brucella sp.*;
  - preparing, controlling and supplying International Standard sera and other reference reagents to the NRLs (to standardise tests and reagents in the MS);
  - validating reference reagents (antigens and National Standard sera);
  - building up and maintaining a sera bank, a collection of *Brucella sp.*, and a database of strains isolated across the EU;
  - organising periodical comparative tests of diagnostic procedures at the EU level and operating laboratory proficiency tests of NRLs;
  - collecting and collating data and information on the methods of diagnosis used and the results of tests carried out in the EU;

- characterizing *Brucella sp.* by the most up-to-date methods available to allow for greater understanding of the epidemiology of brucellosis;
- keeping abreast of development in brucellosis surveillance, epidemiology and prevention throughout the world;
- acquiring thorough knowledge of the preparation and use of the products of veterinarian immunology used to eradicate and to control brucellosis including the evaluation of vaccines.

(ii) Facilitation of harmonization of techniques throughout the EU, in particular specifying standard test methodologies

(iii) Organisation of workshops for the benefit of NRLs, including training of experts, from the MS and, as appropriate, from non-member countries, in new analytical methodologies

(iv) Provision of technical assistance to the Commission, concerning in particular the standardisation of analytical methods and their implementation.

(v) Performance of research activities and whenever possible co-ordination of research activities directed towards improving control and eradication of brucellosis, specifically by:

- carrying out or collaborating with NRLs in carrying out test validation trials;
- providing scientific advise to the Commission services and collecting information and reports associated with the activities of the CRL.

This is a long list of missions that obviously needs to be prioritised. However, these missions fall within an already existing and well-organised network including the EC, the Task Force, the FVO, EDQM, EFSA and all MS NRLs. Thus, there is strong hope for more coordinated and integrated organisation for the standardisation of diagnosis, eradication/surveillance programmes, and improvement of epidemiological knowledge of the disease.

The official authority given to the CRL for coordinating the NRLs network and the provision of adequate funding to the CRL for the achievement of its missions, two conditions officially foreseen by Commission Regulations (EC) N° 1754/2006 and (EC) N° 882/2004 represent the key to success for this program. They provide preliminary assurance that, on one hand, the CRL will be able to make the necessary efforts to set up EU-wide standards for testing, routine procedures and reliable testing methods and, on the other hand, that the MS NRLs will make similar efforts to fulfil common procedures and protocols.

References

- Council Directive 64/432/EEC of 26 June 1964 on health problems affecting intra-Community trade in bovine animals and swine: http://europa.eu/scadplus/leg/en/lvb/l12004.htm.
- Council Directive 91/68/EEC of 28 January 1991 on animal health conditions governing intra-Commu-

nity trade in ovine and caprine animals: http://europa.eu/scadplus/leg/en/lvb/l12005.htm.

- Commission Decision of 28 October 2005 amending Decision 93/52/EEC as regards the declaration that the province of Grosseto in the Region of Toscana in Italy is free of brucellosis (*B. melitensis*) and Decision 2003/467/EC as regards the declaration that France is free of bovine brucellosis: http://eur-lex.europa.eu/LexUriServ/site/en/oj/2005/l_288/l_28820051029en00560058.pdf.
- White Paper on Food Safety - Commission of the European Communities - Brussels, 12 January 2000 - COM (1999) 719 final: http://ec.europa.eu/food/food/intro/white_paper_en.htm.
- Council Decision 90/424/EEC of 26 June 1990 on expenditure in the veterinary field: http://eur-lex.europa.eu/LexUriServ/LexUriServ.do?uri=CELEX:31990D0424:EN:HTML
- Regulation (EC) No 882/2004 of the European Parliament and of the Council of 29 April 2004 on official controls performed to ensure the verification of compliance with feed and food law, animal health and animal welfare rules:
http://eur-lex.europa.eu/LexUriServ/site/en/oj/2004/l_191/l_19120040528en00010052.pdf.
- Commission Regulation (EC) No 776/2006 of 23 May 2006 amending Annex VII to Regulation (EC) No 882/2004 of the European Parliament and of the Council as regards Community reference laboratories: http://eur-lex.europa.eu/LexUriServ/site/en/oj/2006/l_136/l_13620060524en00030008.pdf.
- Commission Decision of 19 September 2006 as regards financial aid from the Community for the year 2006 for the Community reference laboratories for brucellosis and foot-and-mouth disease: http://eur-lex.europa.eu/LexUriServ/site/en/oj/2006/l_258/l_25820060921en00050006.pdf.
- Commission Regulation (EC) No 1754/2006 of 28 November 2006 laying down detailed rules for the granting of Community financial assistance to Community reference laboratories for feed and food and the animal health sector:
http://eur-lex.europa.eu/LexUriServ/site/en/oj/2006/l_331/l_33120061129en00080018.pdf.
- Commission Decision of 5 December 2006 amending Council Directives 64/432/EEC, 90/539/EEC, 92/35/EEC, 92/119/EEC, 93/53/EEC, 95/70/EC, 2000/75/EC, 2001/89/EC, 2002/60/EC and Decision 2001/618/EC as regards lists of national reference laboratories and State institutes: http://eur-lex.europa.eu/LexUriServ/site/en/oj/2006/l_346/l_34620061209en00410058.pdf.
- European Task Force for monitoring disease eradication in the member states: http://ec.europa.eu/food/animal/diseases/eradication/taskforce_en.htm.
- COST Action 845 “Brucellosis in man and animals”: http://www.brucella.org/cost/costhome.asp.

Dr. B. Garin-Bastuji, European Community & OIE/FAO Reference Laboratory for brucellosis, Agence française de Sécurité Sanitaire des Aliments, 23 avenue du Général de Gaulle, 94706, Maisons-Alfort, France. E-mail: b.garin-bastuji@afssa.fr

Lombard M, Dodet B, the OIE Scientific & Technical Department (eds): First International Conference of the OIE Reference Laboratories and Collaborating Centres. Dev Biol (Basel). Basel, Karger, 2007, vol 128, pp 139-143.

The Role of Diagnostic Laboratories in Support of Animal Disease Surveillance Systems

C. Zepeda

OIE Collaborating Center for Animal Disease Surveillance and Risk Analysis, Centers for Epidemiology and Animal Health USDA-APHIS-VS, Fort Collins, Colorado, USA

Key words: Animal disease, surveillance system, diagnostic laboratories, OIE

Summary: Diagnostic laboratories are an essential component of animal disease surveillance systems. To understand the occurrence of disease in populations, surveillance systems rely on random or targeted surveys using three approaches: clinical, serological and virological surveillance. Clinical surveillance is the basis for early detection of disease and is usually centered on the detection of syndromes and clinical findings requiring confirmation by diagnostic laboratories. Although most of the tests applied usually perform to an acceptable standard, several have not been properly validated in terms of their diagnostic sensitivity and specificity. Sensitivity and specificity estimates can vary according to local conditions and, ideally, should be determined by national laboratories where the tests are to be applied. The importance of sensitivity and specificity estimates in the design and interpretation of statistically based surveys and risk analysis is fundamental to establish appropriate disease control and prevention strategies.

The World Organisation for Animal Health's (OIE) network of reference laboratories acts as centers of expertise for the diagnosis of OIE listed diseases and have a role in promoting the validation of OIE prescribed tests for international trade. This paper discusses the importance of the epidemiological evaluation of diagnostic tests and the role of the OIE Reference Laboratories and Collaborating Centres in this process.

Introduction

The role of disease surveillance systems is to transform field and laboratory data into information that can be used to base decisions related to the control and eradication of animal diseases. Specifically, surveillance systems collect, analyze and interpret data to:

- determine the distribution of diseases in time and space and,
- determine the presence or absence of disease.

The growth of international trade of animals and products has placed increased demands on surveillance systems to provide the necessary assurances that trade can be conducted safely. Disease freedom claims need to be supported by surveillance information to provide an adequate confidence level of the absence of disease. Once disease freedom has been recognized, on-going surveillance is required to maintain the status of a disease free country or zone. Compartmentalization, which is defined as the recognition of animal populations of different health status based on management and biosecurity [1], also requires increased surveillance both within the compartment as well as in the surrounding areas.

All the above activities require large numbers of samples to be collected, analyzed and interpreted. In addition, in the past few years, the world has seen major outbreaks of animal diseases like foot-and-mouth disease and avian influenza, placing considerable pressure on veterinary services and highlighting the importance of disease surveillance systems and diagnostic laboratories.

Surveillance Approaches

Animal disease surveillance systems combine several approaches to achieve their objective [2]. Random surveillance is based on statistically based surveys to determine the disease status of an animal population. Frequently, depending on the survey design, a large number of samples is required to achieve an acceptable level of confidence.

In order to reduce the burden on surveillance systems, a second approach has been applied with the objective of reducing the number of samples while maintaining an adequate confidence level. This can be achieved by targeting surveillance efforts to high-risk groups in the population. This approach is particularly useful for surveys designed to determine the presence or absence of disease, as it allows the use of higher design prevalence values, leading to a reduced number of samples required. Obviously, animal disease surveillance is dependent on diagnostic test results and competent laboratories to perform these tests.

Due to practical reasons and resource limitations, not all OIE listed diseases can be included under active surveillance programmes. Therefore, a key question for any surveillance programme is to define the diseases that will be included in the system and the intensity of surveillance that will be used. The choice of disease priorities should be based on [3]:

- the potential to have a public health impact,
- diseases that significantly affect production and
- diseases that limit international trade.

For diseases under a control and eradication programme, an active surveillance strategy using clinical, serological and virological methods is required to confirm the presence of the disease and plan actions to reduce the incidence. For other diseases, a passive surveillance system based on the notification of syndromes is an efficient approach to detect cases.

The Role of Diagnostic Laboratories in Support of Surveillance Systems

OIE Reference Laboratories have a role in validating diagnostic tests. The process of validation includes among other things, determining their diagnostic sensitivity (Se) and specificity (Sp). Diagnostic sensitivity and specificity estimates are extremely relevant for the planning and interpretation of surveillance activities.

An important epidemiological measure used in the interpretation of diagnostic tests results is the predictive value, which can be defined as the confidence that can be placed on diagnostic test results. Positive and negative predictive values (PPV and NPV) are dependent on Se, Sp and the prevalence [4]. Positive predictive values drop sharply as the prevalence decreases. This means that a test that yields an acceptable PPV at the beginning of a disease control programme, when the prevalence is high, will have an extremely low PPV when approaching eradication. The reverse occurs with the negative predictive value. Appropriate diagnostic strategies need to be adapted depending on the epidemiological situation. The effect of these adjustments can only be measured if accurate Se and Sp estimates are available.

Risk analysis results also depend on accurate Se and Sp estimates. When assessing the probability of introducing a disease agent through the importation of live animals or when interpreting the results of surveys conducted to declare a country, zone or compartment as free from disease, Se and Sp estimates for OIE prescribed tests are essential. Relatively small variations in these estimates can lead to very different results and potentially to risky decisions.

A common problem for the epidemiologist is to find good Se and Sp estimates in the scientific literature. Validation studies are difficult to perform. Although specificity estimates are, in principle, relatively simple to perform, obtaining samples from known non-infected animals may prove difficult. Determining Se values is more complex, as it requires samples from known infected animals, and laboratory conditions to conduct such studies are limited and expensive. This frequently means that very few animals are used in these studies, leading to very wide confidence intervals around Se estimates.

Another problem is that new tests are frequently validated against a gold standard that, in turn, has not been properly validated in terms of its Se and Sp and therefore it is not possible to easily determine the Se and Sp of the new test.

In recognition of these problems, a special issue of Preventive Veterinary Medicine [5] was devoted to the validation of diagnostic tests. Some of the issues addressed in this issue include [6]:

- Se and Sp estimates when the gold standard is imperfect
- Se and Sp estimates when there is no gold standard

- Pool interpretation of test results
- Series or parallel interpretation of correlated tests
- Herd level sensitivity and specificity estimates
- Declaration of disease free zones and compartments

OIE Reference Laboratories promote and support national laboratories in the implementation of diagnostic tests. Ideally, Se and Sp estimates should be determined at the national level as local production conditions such as breed, nutrition and concomitant infections might lead to changes in the performance of the test.

The large number of samples required by different surveillance strategies and purposes requires the development of methodologies that have a quick turn around time and can process large amounts of samples.

In the recent past, several large scale outbreaks have prompted the destruction of millions of animals (foot-and-mouth disease, classical swine fever, avian influenza) [7-9]. The public reaction to such policies is likely to force a change in the way we deal with disease incursions. Increasingly, disease control and eradication strategies are considering vaccination as an option in addition to or in substitution of stamping-out. The most important limiting factor for the use of vaccines in this context is the differentiation between infected and vaccinated animals.

Several vaccines and their corresponding diagnostic tests allowing this differentiation have already been developed for several diseases [10]. The current growth of biotechnology is helping drive an expansion in the number of tests, vaccines and diagnostic strategies that will allow a more widespread use of vaccines in eradication programmes. Similarly, several multiplex tests that are able to test simultaneously for several disease agents have been developed.

OIE Reference Laboratories play a central role in the development, validation and harmonization of new diagnostic tests and vaccines.

Conclusion: Collaboration between OIE Reference Laboratories and Collaborating Centres

Diagnostic processes should not be conducted in isolation; they need to respond to the epidemiologic reality in the field. Therefore, a close and permanent communication between epidemiologists and scientists working in diagnostic laboratories as well as those involved in the development of new tests and vaccines should be maintained. This interaction should occur at the international level between OIE Reference Laboratories and Collaborating Centres, in particular those dealing with epidemiology, as well as at the national level between the national laboratories and the disease surveillance system.

The OIE, through its network of Reference Laboratories and Collaborating Centres, should promote studies that provide more accurate estimates for the diagnostic sensitivity and specificity of the prescribed tests for international trade.

This will allow epidemiologists and decision-makers to reduce the uncertainty in decisions related to the recognition of disease freedom and import risk assessments.

References

1. Scott A, Zepeda C, Garber L, Smith J, Swayne D, Rhorer A, et al: Concept of compartmentalisation. OIE Sci Tech Rev, 2007, (in press).
2. OIE: Terrestrial Animal Health Code. General Guidelines for Surveillance. World Organization for Animal Health, Paris, 2006, (www.oie.int).
3. Zepeda C, Salman MD: Planning survey, surveillance and monitoring systems – roles and requirements. In Animal Disease Surveillance and Survey Systems, Methods and Applications. MD Salman Ed. Iowa State University Press, Ames, 2003.
4. Dohoo I, Martin W Stryhn H: Veterinary Epidemiologic Research. AVC Inc., Charlottetown, Prince Edward Island, 2003.
5. Gardner I, Greiner M: Preface. Prev Vet Med 2000;45:1.
6. Gardner I, Greiner M: Epidemiologic issues in the validation of veterinary diagnostic tests. Prev Vet Med 2000;45:1:3-22.
7. OIE: World Animal Health in 1998. World Organization for Animal Health, Paris, 1999, (www.oie.int).
8. OIE: World Animal Health in 2001. World Organization for Animal Health, Paris, 2002. (www.oie.int).
9. OIE: World Animal Health in 2003. World Organization for Animal Health, Paris, 2004, (www.oie.int).
10. OIE: Manual of Standards for Diagnostic Tests and Vaccines. World Organization for Animal Health, Paris, 2006, (www.oie.int).

Dr. Cristóbal Zepeda, OIE Collaborating Center for Animal Disease Surveillance and Risk Analysis, Centers for Epidemiology and Animal Health USDA-APHIS-VS. 2150 Centre Ave. Fort Collins, Colorado, 80525, USA.
E-mail: cristobal.zepeda@aphis.usda.gov

Lombard M, Dodet B, the OIE Scientific & Technical Department (eds): First International Conference of the OIE Reference Laboratories and Collaborating Centres. Dev Biol (Basel). Basel, Karger, 2007, vol 128, p 145.

Session VI

General Recommendations

Lombard M, Dodet B, the OIE Scientific & Technical Department (eds): First International Conference of the OIE Reference Laboratories and Collaborating Centres. Dev Biol (Basel). Basel, Karger, 2007, vol 128, pp 147-150.

Resolution: Strengthening the Networking between OIE Reference Laboratories, OIE Collaborating Centres and National Veterinary Laboratories to Enhance the Even Global Spread of Veterinary Scientific Knowledge and Expertise

CONSIDERING THAT

1. The global network of 180 Reference Laboratories and Collaborating Centres of the OIE is a unique key source of expert support and information in setting science-based international standards, guidelines and recommendations for the prevention, detection and control of animal diseases, including zoonoses, animal welfare and animal production food safety,
2. The OIE network of Reference Laboratories and Collaborating Centres is a base for the worldwide veterinary scientific community,
3. Good governance is strongly linked with capacity-building programmes at the global, regional and national levels,
4. The OIE is committed to promoting global animal health by assisting developing and in-transition countries to acquire and apply good governance to enable compliance with the application of the guidelines, recommendations and international standards of the OIE,
5. More than 72% of the 168 Member Countries of the OIE are developing and in-transition countries with various scientific capacities and means of access to scientific expertise,
6. There is an urgent need to establish more OIE Reference Laboratories and OIE Collaborating Centres in developing and in-transition countries to strengthen the veterinary scientific communities in these countries so that they can assist in the preparation of OIE guidelines, recommendations and

standards, and face new threats linked to globalisation and climatic changes,

7. Twinning between OIE Reference Laboratories and OIE Collaborating Centres with candidate laboratories within developing and in-transition countries would lead to a more even geographical distribution of expertise, and the establishment of more OIE Reference Laboratories and OIE Collaborating Centres would facilitate access to expertise for, notably, the rapid detection and diagnosis of diseases,
8. The International Committee of the OIE during the 70th General Session adopted Resolution No. XXI that identified The Role of OIE Reference Laboratories and Collaborating Centres in Capacity Building for Developing Countries,
9. There is a need to strengthen cooperation and the exchange of knowledge and expertise within the network of OIE Reference Laboratories and OIE Collaborating Centres and to increase access to and use of this knowledge by the national veterinary laboratories and the Veterinary Services of all OIE Member Countries,
10. There is a continuous need to update approaches to the setting of standards,
11. The cost of and current regulatory limitations relating to the transport of specimens, especially shipment by air, between laboratories, could have a negative impact on the rapid detection and diagnosis of animal diseases,
12. The exchange and sharing of information through the OIE World Animal Health Information System (WAHIS), is essential for enhancing transparency and collaboration between laboratories,
13. For OIE and FAO Reference Laboratories, exchange of information on laboratory testing results is integral to the joint FAO/OIE/WHO GLEWS (Global Early Warning System) platform,
14. There is a need for OIE Reference Laboratories and OIE Collaborating Centres to meet annually or bi-annually to strengthen the network, to exchange information and experiences, and to ensure the consistent application of OIE guidelines, recommendations and standards.

The First International Conference of OIE Reference Laboratories and Collaborating Centres Recommends the Following:

1. The OIE Reference Laboratories and OIE Collaborating Centres are encouraged to give priority to capacity-building programmes at the global, regional and national levels under the auspices of the OIE, in collaboration with FAO and relevant national and regional organisations.
2. Networks of OIE Reference Laboratories and OIE Collaborating Centres

should be developed with the objective of harmonising and exchanging data, information and reference and diagnostic material to improve world-wide disease surveillance and control and the implementation of OIE objectives and strategies.

3. The relevant information generated by the OIE Reference Laboratory and OIE Collaborating Centre network should be integrated in the OIE on-line WAHIS information system.
4. The OIE Reference Laboratories and OIE Collaborating Centres are encouraged to actively exchange data, information and reference material within the given limitations of individual property rights and within the opportunities offered through the OIE WAHIS information system.
5. The networks of OIE Reference Laboratories and OIE Collaborating Centres should strive to achieve full compliance with and actively pursue opportunities to update approaches to the setting and uniform application of OIE guidelines, recommendations and standards.
6. The networks of OIE Reference Laboratories and OIE Collaborating Centres will be open to accepting collaboration with FAO Laboratories and Collaborating Centres and those from relevant regional and national organisations, under the auspices of the OIE.
7. The Director General of the OIE should pursue negotiations with donors to secure funding for collaborative programmes by OIE Reference Laboratories and OIE Collaborating Centres to establish and maintain international reference materials for OIE-listed diseases.
8. The Director General of the OIE should further pursue negotiations with the international organisations responsible for establishing relevant international requirements for the shipment of specimens.
9. The OIE Reference Laboratories and OIE Collaborating Centres shall be encouraged to enter into agreements to twin with candidate laboratories in developing and in-transition countries.
10. The OIE Reference Laboratories and OIE Collaborating Centres shall be encouraged to make every effort to share scientific knowledge and skills with laboratories and centres in developing and in-transition countries and provide relevant training, in particular in the development and implementation of rapid, robust and inexpensive diagnostic tests and in vaccine development so that disease control programmes can be improved.
11. The Director General of the OIE should pursue negotiations with donors to secure funding for priority twinning agreements between OIE Reference Laboratories and OIE Collaborating Centres and candidate laboratories and collaborating centres in developing and in-transition countries in support of the commitment of the OIE to establishing scientific capacity building in

developing and in-transition countries.

12. The International Committee of the OIE should reconsider the decision related to the obligation of OIE Reference Laboratories to report positive results directly to the OIE Central Bureau.
13. The issue of whether or not distinct genotypes of OIE listed diseases should be listed as such and have individual reporting requirements should be discussed in a wider forum at the next meeting of OIE Reference Laboratories.
14. OIE Reference Laboratories should be requested by the OIE to take the lead in developing and harmonising methods for differentiation of genotypes.
15. The Director General of the OIE should create the opportunity to host a similar Conference in 2008 for OIE Reference Laboratories and Collaborating Centres and those from FAO and other partners and including those laboratories participating in twinning agreements.

Lombard M, Dodet B, the OIE Scientific & Technical Department (eds): First International Conference of the OIE Reference Laboratories and Collaborating Centres. Dev Biol (Basel). Basel, Karger, 2007, vol 128, p 151.

Poster Session

Abstracts

Lombard M, Dodet B, the OIE Scientific & Technical Department (eds): First International Conference of the OIE Reference Laboratories and Collaborating Centres. Dev Biol (Basel). Basel, Karger, 2007, vol 128, p 153.

Quality Control and Stability Tests of *Brucella Abortus* Strain 19 Vaccine Used in Argentina

A.M. Nicola, S. Elena

OIE Reference Laboratory for Brucellosis, National Diagnostic Laboratory of Animal Health Services (DILAB-SENASA), Buenos Aires, Argentina

In Argentina, the National Direction of Laboratories and Technical Control of Animal Health Service (DILAB-SENASA) conducts a national assay on biological products (vaccines, immune sera and diagnostic reagents). In 1993, quality control of the *Brucella abortus* strain 19 vaccine used in Argentina was established with a standard assay used on each batch and minimum standards: sterility, purity and identity, pH, viable organisms (15 to 30 CFU/109/doses) and smoothness (not more than 5% dissociation permitted). All batches of Brucellosis vaccine used are manufactured in Argentina by seven private laboratories and they must be approved and licensed by SENASA.

In Argentina, about 230 batches of *Brucella abortus* S19 are approved per year (nearly 10,000,000 doses); no more than 20 batches, have been rejected in the last five years. The reasons for rejection were non conformity with regard to sterility, dissociation, the number of viable bacteria and loss of the vacuum seal.

To establish the validity of the expiry date, the loss of viability and the appearance of dissociation due to temperature stress of five batches from different manufacturers were re-tested at the approximate expiration date, and stability studies done incubating the reconstituted vaccine at 37°C for 1 week. The results of this study showed that all batches conserved the identity of the strain and five batches conserved the number of viable bacteria obtained one year prior to product approval. Two batches lost viability, with over 60% of vaccines under the minimum CFU required for product approval. Dissociation was not observed. All the vaccines were stored under the same conditions. The difference between batches observed in stability studies could be due to the master seed used in the production of each batch.

Brucellosis vaccination is mandatory in Argentina and immunization of animals with high quality vaccines is the primary means of control for this important zoonotic disease.

Lombard M, Dodet B, the OIE Scientific & Technical Department (eds): First International Conference of the OIE Reference Laboratories and Collaborating Centres. Dev Biol (Basel). Basel, Karger, 2007, vol 128, p 154.

Brucellosis in Marine Mammals: Veterinary Laboratories Agency Involvement

E. Stubberfield, C. Dawson, S. Brew, L. Perrett, J. Stack

Veterinary Laboratories Agency, Woodham Lane, Addlestone, Surrey, UK

Brucellosis is an important zoonotic disease usually associated with cattle, sheep, goats and pigs. In 1994 *Brucella* was isolated from four Common seals (*Phoca vitulina*), two Harbour porpoises *(Phocoena phocoena)* and one Common dolphin (*Delphini delphinus*) in Scotland and a captive Bottlenose dolphin (*Tursiops truncatus*) in the USA. These isolates were found to be morphologically and molecularly different from the terrestrial strains and have been proposed the names *Brucella maris* or *B. cetacea* and *B. pinnipedia*, based on host specificity.

There is serological evidence of brucellosis in marine mammal species around the world, however, the organism has only been isolated in the northern hemisphere. *Brucella* strains that have been isolated from sea mammals have been shown to be zoonotic by four reported cases of infection in humans. The first human case was reported in 1999, a laboratory worker acquired infection in the UK, and three further cases are believed to be linked to the consumption of raw sea food.

The Veterinary Laboratories Agency (VLA) has been involved in the examination of marine mammals both serologically and with the culture and molecular typing of *Brucella* isolated from various tissues. Since the first isolations in 1994, the VLA has tested 151 isolates from a variety of marine mammals, primarily from British waters, but also from the USA, Canada and Northern Europe. We have tested over 3,900 serological samples from around the world, from both captive and wild animals, by three ELISAs.

Lombard M, Dodet B, the OIE Scientific & Technical Department (eds): First International Conference of the OIE Reference Laboratories and Collaborating Centres. Dev Biol (Basel). Basel, Karger, 2007, vol 128, p 155.

Veterinary Laboratories Quality Evaluation of Brucellosis Diagnosis in Argentina

A.M. Nicola, Z. Alonso, S. Elena, J. Esteves Madero

Reference Laboratory OIE for Brucellosis, National Direction of Laboratories and Technical, Control of Animal Health Services (DILAB-SENASA), Buenos Aires, Argentina

Veterinary laboratories for infectious diseases are a vital component of the infrastructure that provides animal health services.

In Argentina, the National Direction of Laboratories and Technical Control of Animal Health Services (DILAB-SENASA) initiated, in 1995, a diagnostic laboratory network and developed a procedure to approve state and private laboratories to perform certain diagnostic tests.

The Brucellosis Department of the DILAB-SENASA has established a Quality System to ensure the validity of results of the laboratories conducting brucellosis diagnosis following the standard procedures as outlined by SENASA, ISO 17025 and the OIE, in support of Brucellosis Control or Eradication programmes, for export and interstate movement.

In Argentina, there are now 235 approved state and private laboratories approved for the brucellosis serological test (BPAT, SAT, 2-Mercaptoethanol), four laboratories for CF testing and, as of 2006, three private laboratories for FPA testing and one state laboratory for ELISA testing. The Brucellosis Laboratory Quality Program addresses administrative information, laboratory facilities, maintenance and calibration of equipment, handling of test results, laboratory staff training, quality assurance procedures, use of SENASA-licensed reagents, proficiency testing and reporting of all test results to animal health officials.

The objective of the Brucellosis Department is to conduct proficiency testing once per year using 20 blind serum samples (strong positive, weak positive and negative by replicate) and a positive and negative sera control. Z-scores are used to evaluate laboratory performance in the quantitative test for each sample.

In total, 90 laboratories participated in 2006, conducting a brucellosis test panel by BPAT, SAT and 2-Mercapatoethanol. The results showed that participating laboratories have an acceptable agreement of 87%. The Laboratories conducting CF, FPA and ELISA have 100% agreement.

The brucellosis agglutination diagnosis test contains an element of subjectivity in the interpretation of results, and this renders both internal and external assurance difficult. Results demonstrated that several laboratories still need to improve their laboratories practices, and should also regularly check the calibration of their equipment.

Lombard M, Dodet B, the OIE Scientific & Technical Department (eds): First International Conference of the OIE Reference Laboratories and Collaborating Centres. Dev Biol (Basel). Basel, Karger, 2007, vol 128, p 157.

Identification of *Brucella* Spp. by PCR (Polymerase Chain Reaction) from Isolations and Clinical Samples

M.D. Trangoni, M.E. Cerón, A.M. Nicola[2], M.A. Fiorentino[1], B. Alonso[2], S. Elena[2], O. Rossetti, S.L. Cravero

Instituto de Biotecnología, CICVyA, INTA Cautelar, [1] Grupo de Sanidad Animal, EEA INTA Balcarce, República Argentina, [2] Dirección de Laboratorios y Control Técnico, Servicio Nacional de Sanidad y Calidad Agroalimentaria, Reference Laboratory OIE for Brucellosis, (DILAB-SENASA), República Argentina

Brucellosis is a zoonosis that is found worldwide, is extensively diffused, and affects wild and domestic mammals. This illness was characterized over 100 years ago and significant advances in control and diagnostic methods have been made. However, laborious methods for characterization of the genus, species and biotype are still being used. The identification of *Brucella spp.* by PCR allows for a diagnosis to be made with dead bacteria, unlike bacteriological methods. The availability and comparison of genomic sequences of the most pathogenic species for humans (*B. abortus, B. melitensis* and *B. suis*) allows for the identification of nucleotide sequences characteristic of each species.

The objective of this work was the application of the PCR technique for the identification of genus and species of *Brucella* from bacterial isolation and clinical field samples. The present study involves: reference strains *B. abortus* S 2308, *B. melitensis* 16M, *B. suis* 1330, *B. ovis* Reo 198, and *B. canis* RM6/66; field isolated strains *B. abortus*, *B. melitensis*, *B. suis*, *B. ovis*, and *B. canis*; vaccine strains RB 51 and S19; and milk, bovine sera and abdominal cavity and bovine fetus abomasums fluid collected from naturally infected animals. DNA extraction was carried out according to standard laboratory protocol and was done with commercial kits. The primers utilized (B4-B5, I-S711-AB) have been previously described by other authors. Primers were designed, based on the comparative analysis of the three genomes available in GenBank, to amplify specific regions (Ab1F-Ab1R, Me1F-Me1R, Su1F-Su1R, XAb1F-XAb1R, and XSu1F-

XSu1R). This assay identified *Brucella* spp. DNA from clinical samples and allowed for the typing of reference strains. It is necessary to validate these tests. It is possible to design a multiplex PCR with which the *Brucella* species could be defined in a single reaction. The differentiation between *B. suis* and *B. canis* could be done using oligonucleotides specific for the ORF BMEI1436, absent in *B. canis*. This test can be used to complement existing tests in the diagnosis of brucellosis.

Lombard M, Dodet B, the OIE Scientific & Technical Department (eds): First International Conference of the OIE Reference Laboratories and Collaborating Centres. Dev Biol (Basel). Basel, Karger, 2007, vol 128, p 159.

Immunofluorescence and Serodiagnosis of *Leptospira* in Farm Animals

B. Brihuega, G. Romero, C. Auteri, L. Samartino

INTA, CICVyA-Pathobiology-Bacteriology Unit, Leptospirosis Division, Buenos Aires, Argentina

Leptospirosis is a worldwide zoonotic disease caused by *Leptospira interrogans*. In animals, leptospirosis results in severe economic losses mainly due to reproductive disorders like abortion, stillbirth, congenital infection and infertility. In cattle, abortions are more frequent in infections beginning during the last third of the gestation period. Leptospirosis in horses is marked by fever and anorexia, usually not recognized. In pregnant mares, abortions are also present. Leptospirosis affects swine of all ages but is most frequently recognized through abortions. We present here the results of a survey done in the last 2 years in bovines, equines and porcines.

We studied 2,742 samples (2,394 bovine, 223 equine and 125 porcine) obtained from the provinces of Buenos Aires, Cordoba, Entre Rios, La Pampa and Santa Fe in the Argentine Republic. All samples were evaluated with the microaglutination test (MAT) which is the international serological reference test. Additionally, the immunofluorescence (IF) test was performed on 337 bovine and 65 equine samples.

MAT showed 61.50% of the bovine samples to be positive, as well as 59.64% of the equine and 57% of porcine samples. Predominant serovars varied depending on the animal species: bovine serovars were *L. hardjo* (43%), *L. wolffi* (41%), *L. pomona* (13%) and *L. tarassovi* (12%); equine serovars were *L. pomona* (38%), *L. ballum* (26.9%), *L. wolffi* (17%), *L. icterohaemorrhagiae* (5.8%), *L. tarassovi* (4%) and *L. grippotyphosa* (1.34%); and porcine serovars were *L. icterohaemorrhagiae* (84%), *L. pomona* (30 %), *L. grippotyphosa* (25%), *L. tarassovi* (4%) and *L. canicola* (2%). With the IF test, 28.78% of bovine samples were found to be positive as well 50.76% of the equine samples.

According to these results, there has been an increase in leptospirosis seropositive animals as compared to previous studies. Interestingly, a higher percentage of bovine samples were positive for serovar *L. hardjo*, when in almost all previous studies it was shown that *L. pomona* was the serovar most frequently detected in bovine leptospirosis in Argentina. A remarkable number of porcine samples were positive to *L. icterohaemorragiae* whereas *L. pomona* used to be the predominant serovar in these animals. The serological evidence reveals that the incidence of this zoonosis has increased in the last two years, and the high number of positive animals in this study agrees with an increase of the human cases of leptospirosis. It is recommended that there be an increase in the control and prevention of the disease in animals, as well as in humans, through rodent and garbage control and public education.

Lombard M, Dodet B, the OIE Scientific & Technical Department (eds): First International Conference of the OIE Reference Laboratories and Collaborating Centres. Dev Biol (Basel). Basel, Karger, 2007, vol 128, p 161.

Detection of *Mycobacterium bovis* Infection in Farm and Companion Animals in Great Britain

K.L. Jahans, D.R. Worth, J. Brown, J.M. Gunn, M. Okker, E. Wood, L. Potts

Veterinary Laboratories Agency, New Haw, Addlestone, Surrey, UK

The number of incidents of tuberculosis in cattle in Great Britain has increased by an average of 16% each year since the mid 1980s.

Lombard M, Dodet B, the OIE Scientific & Technical Department (eds): First International Conference of the OIE Reference Laboratories and Collaborating Centres. Dev Biol (Basel). Basel, Karger, 2007, vol 128, p 162.

Phylogenetic Analysis and Epidemiological Tracing of Viral Haemorrhagic Septicaemia Virus (VHSV)

N. Jørgen Olesen, S. Madsen, K. Einer-Jensen, H.F. Skall, N. Lorenzen

Danish Institute for Food and Veterinary Research, OIE Reference Laboratory for VHS, Århus, Denmark

Viral haemorrhagic septicaemia (VHS) caused by the rhabdovirus VHSV is one of the most important viral diseases in European rainbow trout farming. Until 1989, the virus was mainly isolated from freshwater salmonids, especially rainbow trout. However, in the last 17 years, it has also been isolated from an increasing number of free-living marine fish species. The marine and fresh water isolates are serologically indistinguishable, but differ with respect to pathogenicity in the host species; in general, marine VHSV isolates do not induce disease in rainbow trout. Recent studies have shown that most diversity is within the marine virus group. Thus far, attempts to identify a marker that may link genetic and pathogenic characteristics has been unsuccessful.

A large amount of VHSV sequence data has been generated since 1998, and parallel phylogenetic analysis using various genomic regions from a fixed virus panel reveal that the overall genotype of VHSV is identical regardless of which target region is selected (Einer-Jensen et al. 2005). The choice, and especially the length of the target region, however, influenced the ability to characterise the isolates into genetic sub-lineages; this observation should be kept in mind during epidemiological analyses. VHSV isolates can be divided into 4 genogroups: I, II, III and IV. Genogroup IV and I are further divided into subgroups. The rainbow trout pathogenic isolates primarily cluster in Ib. The advantages and disadvantages of differentiating between VHSV genotypes upon OIE notification will be discussed. On several occasions, the existing sequence data have proven to be informative when used for epidemiological tracing of new VHSV outbreaks. Case stories from 2006 as well as plans regarding the generation and management of additional sequence information will also be discussed.

Lombard M, Dodet B, the OIE Scientific & Technical Department (eds): First International Conference of the OIE Reference Laboratories and Collaborating Centres. Dev Biol (Basel). Basel, Karger, 2007, vol 128, p 163.

An International Standard for *Brucella Melitensis*: The Preliminary Work

A. Taylor, L. Duncombe, J. Stack, J. Mcgiven

Department of Statutory and Exotic Bacterial Disease, Veterinary Laboratories Agency, Addlestone, United Kingdom

At the request of the OIE, The Veterinary Laboratories Agency prepared a candidate International Standard serum against *Brucella melitensis* (ISaBmS) for use in small ruminants. The aim was to create an OIE standard, which would be used to control species specific tests in sheep and goats (eg iELISA) in the same way that laboratories currently use the OIEISS and ELISA standards in bovine brucellosis. The ISaBms standard could potentially be used for other tests which diagnose sheep and goat brucellosis (Rose Bengal, Complement Fixation, Fluorescence Polarisation Assays and cELISA) that are currently standardised against the bovine standard. Serum for this was obtained from a number of naturally infected and experimentally infected goats. This serum was pooled, lyophilised and labelled as the proposed *B. melitensis* International Standard serum. Collaborative testing was undertaken in the National Reference Laboratories of Argentina, Belgium, Canada, France, Germany, Greece, Italy, Israel, Portugal, Spain and the United Kingdom in 2006. The laboratories were requested to test a series of dilutions of the standard using various assay types in order to determine which were positive and which were negative and to provide information on the antigen/kits used and any deviations in test procedure from the OIE Manual 2005. These results are tabulated in this poster and are for discussion at this meeting.

Lombard M, Dodet B, the OIE Scientific & Technical Department (eds): First International Conference of the OIE Reference Laboratories and Collaborating Centres. Dev Biol (Basel). Basel, Karger, 2007, vol 128, p 164.

Tasks of an OIE and National Reference Laboratory for the Diagnosis of Enzootic Bovine Leukosis (EBL)

D. Beier

Friedrich-Loeffler-Institut (FLI), Federal Research Institute for Animal Health, Germany

This poster presents the principal tasks of a reference laboratory, using as an example an OIE and National Reference Laboratory for the Diagnosis of enzoonotic bovine Leukosis (EBL).

Task spectrum including research activities:

A. Tasks of an OIE Reference Laboratory for enzootic bovine Leukosis:

- Become an expert centre;
- Production and investigation of reference material;
- Organisation of ring trials;
- Development and investigation of new test methods;
- Collection, analysis and assessment of epidemiological data;
- Hosting advanced training courses for scientists from the member states; and
- Publication and diffusion of information, composition of annual reports for the committee.

B. Tasks of a National Reference Laboratory

- Verification and confirmation of the results obtained by diagnostic laboratories in suspect cases (complex diagnostics for pathogen detection including PCR, taking into account epidemiological questions);
- Establishment of all diagnostic test systems;
- Standardisation of diagnostics, implementation of a (standard) catalogue of methods;
- Preparation and supply of test standards;
- Organisation of regular ring trials on the national level;
- Organisation of advanced training courses for the staff of diagnostic

institutions; and

- Preparation of expert decisions for the federal ministry.

C. Current work

a) Activities in publicly funded research

- Laboratory animal husbandry for sampling of reference materials (blood, serum, milk)
- Investigations on the typical and atypical course of infection and on the pathogenesis of EBL in cattle and sheep
- Development, testing and establishment of molecular test methods, provirus detection (PCR, sequencing), cooperation with national and international Collaborating Centres

b) Other activities

- Characterisation of the BLV provirus subtypes by means of RFLPA
- Production of a BVD free new permanent BLV producing cell line (PO 714)
- Production of an international BLV Reference Standard Serum (E 05)

Lombard M, Dodet B, the OIE Scientific & Technical Department (eds): First International Conference of the OIE Reference Laboratories and Collaborating Centres. Dev Biol (Basel). Basel, Karger, 2007, vol 128, p 166.

Establishment of a Real-Time PCR for the Detection of BLV Infection

C. Gabler[1], *D. Beier*[2], *A. Bondzio*[1]

[1] Institute of Veterinary-Biochemistry, Freie Universität Berlin, Germany
[2] Friedrich-Loeffler-Institut, Institute of Epidemiology, Wusterhausen, Germany

Occurrence of BLV infections with low, transient or missing BLV antibodies and with provirus integration is detectable by env-nested PCR in a block-cycler. The aim of this study was to establish a real-time PCR for the early detection of BLV infection.

PCR was carried out in a block-cycler to verify the authenticity of the amplificate and to obtain the optimal annealing temperature for several primer pairs from the env region. Suitability of these primer combinations was tested in the real-time PCR cycler (RotorGene 3000, Corbett Research) with the SYBR Green I method (Sensemix (dT), Quantace). Samples of BLV infected animals were investigated by real-time PCR. In addition, a nested real-time PCR was developed to achieve higher sensitivity.

The primer pair (forward: 5287-5311; reverse: 5594-5570) proved to be one of the most suitable combinations for real-time PCR (90% efficiency; melting point: 83.75°C). Almost all DNA samples of BLV seropositive cattle tested positive in real-time PCR at CT values between 21 and 32. A very weakly seropositive sample was detected only by nested real-time PCR. Use of dilution series showed that nested real-time PCR is 100 to 1,000 times more sensitive than only one real-time PCR run.

These investigations show that the established real-time PCR is a useful tool to detect BLV infection as well as for the relative estimation of the viral load. For detection of animals with a very low BLV load, nested real-time PCR with a higher sensitivity is recommended.

Lombard M, Dodet B, the OIE Scientific & Technical Department (eds): First International Conference of the OIE Reference Laboratories and Collaborating Centres. Dev Biol (Basel). Basel, Karger, 2007, vol 128, p 167.

Contagious Agalactia and Related Diseases in Sheep and Goats: Selection of Antimicrobials for Treatment Based on in vitro Studies

R.D. Ayling, R.A.J. Nicholas

Veterinary Laboratories Agency, (Weybridge), Addlestone, UK

Contagious agalactia (CA) is an economically important disease of sheep and goats that manifests as mastitis, arthritis, keratoconjunctivitis and sometimes pneumonia and is caused by *Mycoplasma agalactiae*. Other Mycoplasma species namely *M. capricolum* subsp. *capricolum* (Mcc); *M. mycoides* subsp. *mycoides* large colony type (MmmLC); *M. mycoides* subsp. *capri* and *M. putrefaciens* cause similar diseases known as CA syndrome and are more commonly associated with disease in goats, but have also been isolated from sheep. *M. ovipneumoniae* which causes atypical pneumonia and *M. conjunctivae* which causes conjunctivitis are the other mycoplasma species that are routinely associated with disease in sheep and goats.

Killed vaccines are not considered to be efficacious for the prevention of *M. agalactiae* infections although some are produced, including a trivalent vaccine for *M. agalactiae*; MmmLC and Mcc. In the absence of effective vaccines, the selection of antimicrobials for treating affected animals should be based on correct diagnosis and accurate information on their effectiveness. Published MIC data is reviewed and new data shows that some mycoplasma species are more susceptible to the same antimicrobial than others. There is also strong evidence that resistance, as determined in vitro, has developed for many commonly used antimicrobials.

Lombard M, Dodet B, the OIE Scientific & Technical Department (eds): First International Conference of the OIE Reference Laboratories and Collaborating Centres. Dev Biol (Basel). Basel, Karger, 2007, vol 128, p 168.

Comparative Study of the Rose Bengal Test and Competitive Elisa Test in Double-Purpose Herds in the State of Lara, Venezuela

F. Escalona, A. Kowalski, C. Rodríguez-Sallaberry, A. Rodríguez, C. Villasmil, J. Maldonado

Universidad Centroccidental "Lisandro Alvarado", Laboratorio de Diagnóstico Molecular, Laboratorio de Embriología y Endocrinología Molecular, Núcleo Tarabana, Cabudare, Estado Lara, Venezuela

Brucellosis is a zoonotic disease that negatively affects milk production in Latin America. Infections with *Brucella abortus* bacteria observed in human populations are fundamentally caused by the ingestion of non-pasteurized milk or milk products. In Venezuela, 80% of milk production is obtained from cross-bred double-purpose herds, which are generally raised under insufficient sanitary conditions. A significant amount of milk produced is used for production of non-pasteurized cheese for human consumption. In Venezuela, the official screening test for brucellosis is the Rose Bengal test (RBT); the competitive ELISA (cELISA) is only used to definitively confirm results obtained with the RBT. The Universidad Centroccidental "Lisandro Alvarado" performed a trial study to assess the status of brucellosis in the state of Lara, Venezuela. Initially, milk ring tests were performed in 108 farms; 27% tested positive. Serum samples were collected from all animals of farms that tested positive to ring test. A total of 2,821 serum samples were collected and tested with RBT and cELISA assays. The positive index observed was lower with RBT than with cELISA (0.42% vs 4.53%, respectively). Both tests confirm that the positive index in the state of Lara is very low. However, the relative sensitivity of RBT is also very low compared to that of cELISA. This apparent low sensitivity could result in a high number of false negative animals that, consequently, would not be eliminated from the herds. Thus, this would constitute a constant source of infection for other animals and people.

Lombard M, Dodet B, the OIE Scientific & Technical Department (eds): First International Conference of the OIE Reference Laboratories and Collaborating Centres. Dev Biol (Basel). Basel, Karger, 2007, vol 128, p 169.

Validation of a Real-time PCR Assay for Detection of Bovine Herpes Virus 1 in Bovine Semen

J. Wang[1], J. O'Keefe[1], D. Orr[1], L. Loth[1], M. Banks[2], P. Wakeley[2], R. Card[2], D. West[2], G. Ibata[2], C. Van Maanen[3], P. Thoren[4], M. Isaksson[4], P. Kerkhofs[5]

[1]Investigation and Diagnostic Centre-Wallaceville, Biosecurity New Zealand, Ministry of Agriculture and Forestry, Upper Hutt, New Zealand - [2] Veterinary Laboratories Agency, Weybridge, United Kingdom - [3] Animal Health Service, Deventer, The Netherlands - [4] National Veterinary Institute, Uppsala, Sweden - [5] Veterinary and Agrochemical Research Centre, Brussels, Belgium

Bovine herpesvirus-1 is an economically important pathogen in cattle, causing syndromes such as infectious bovine rhinotracheitis (IBR), infectious pustular vulvovaginitis (IPV), infectious balanoposthitis (IPB), abortion and reduced fertility. BoHV-1 is described worldwide. BoHV-1-infected bulls are regarded as life-long carriers of the virus and may potentially shed virus intermittently in their semen. This work was undertaken to find a rapid, sensitive and reliable assay as an alternative prescribed test to the time consuming and cumbersome virus isolation test, to guarantee the safety of semen for artificial insemination, including for international trade.

A real-time PCR was developed and standardized for the detection of BHV-1 from extended bovine semen. Full validation of the PCR assay and DNA extraction method was performed. The real-time PCR is very rapid, specific, highly reproducible and more sensitive than virus isolation, the gold standard. An international inter-laboratory ring trial with the participation of six laboratories was performed in order to further evaluate the reliability of the PCR. The assay was found to be highly reproducible in different real-time PCR amplification-detection platforms by different laboratories. The results of this study indicate that the real-time PCR assay is suitable for the detection of BoHV-1 in extended semen, and would be able to replace slow and laborious virus isolation for the screening testing at artificial insemination centres and for international trade. The final validation dossier compiled from this study has been submitted to the OIE for accreditation as a prescribed test for international trade.

Lombard M, Dodet B, the OIE Scientific & Technical Department (eds): First International Conference of the OIE Reference Laboratories and Collaborating Centres. Dev Biol (Basel). Basel, Karger, 2007, vol 128, p 170.

Resistance Patterns of *Brucella abortus* Biovar 1 Strains S19, RB51 and 2308 useful for Differentiation of Vaccine Strains

K. Leite Miranda[1], F. Padilla Poester[1,2], S. Minharro[1], P. Martins Soares Filho[2], A. Pereira Lage[1]

[1] Escola de Veterinária, Universidade Federal de Minas Gerais, Caixa Postal 567, 30123-970, Belo Horizonte, MG, Brazil, alage@vet.ufmg.br - [2] Laboratório Nacional Agropecuário - Lanagro – Ministério da Agricultura, Pecuária e Abastecimento – Pedro Leopoldo, MG – Brazil

Differentiation among vaccine and field strains is required in areas where vaccination is performed. Also, experiments evaluating vaccines should differentiate vaccine from challenge strains. This study evaluated the growth of *B. abortus* biovar 1 strains S19, RB51 and 2308 on media containing different inhibitor agents. From each bacterial suspension (109 CFU/mL), six tenfold dilutions were prepared. Viable counts were done in duplicate by the drop counting method on tryptose agar (TA), TA in air and TA containing each: i-erythritol (1.0 mg/mL) (TA-e); basic fuchsin (20mg/mL and 80 mg/mL) (TA-bf); thionin (2.5 mg/mL and 10 mg/mL) (TA-t); rifampicin (200mg/mL) (TA-r); and safranin O (200 mg/mL) (TA-s). Plates were incubated at 37°C for up to 96h in 5% CO2. All experiments were repeated three times. A ratio between the log of bacterial count in each treatment and the log of bacterial count on TA in CO2, for each experiment was calculated. No difference in ratio was found from strains grown on TA-bf or TA in air. S19 growth was partially inhibited on TA-e, with a 5-log decrease. S19 and 2308 growth were inhibited on TA-r (7-log decrease); only slight inhibition was shown by RB51. TA-s showed no inhibition for RB51 and 2308, but S19 growth decreased one log. Growth on TA-t differed in the two concentrations used. Ratio for 2308 in TA-t 2.5 mg/mL was near 1.0 and S19 and RB51 showed 0.85 and 0.89 ratios, respectively. Growth of 2308 on TA-t 10mg/mL was not inhibited (ratio equal to 0.98), however, S19 and RB51 were completely inhibited. These results show that vaccine strains could be differentiated from 2308 by TA-t 10mg/mL and confirm that TA-e and TA-r could help in the differentiation among vaccine and field strains of *B. abortus* biovar 1.

Acknowledgments: This study was supported y CNPQ (KLM, APL) and FEP - MVZ

Lombard M, Dodet B, the OIE Scientific & Technical Department (eds): First International Conference of the OIE Reference Laboratories and Collaborating Centres. Dev Biol (Basel). Basel, Karger, 2007, vol 128, p 171.

Validation Assay and Performance Comparison of Serological Tests for Official Use in the Diagnosis of Bovine Brucellosis in Chile

A. Rivera[1], P. Lopetegui[1], S. Quintard[2], J. Soto[3], M. Saldias[2], J. Styles[2], R. Muñoz[3], M. Valverde[3], R. Sepúlveda[4], X. Gonzalez[4], S. Andrade[3], M. Villarroel[3], C. Noli[3], R. Axt[2], J. Smith[5]

[1] División de Protección Pecuaria – SAG Chile - [2] Laboratorio Central SAG Lo Aguirre Chile - [3] Laboratorio Regional SAG Osorno - [4] Oficina SAG Talagante- [5] Oficina SAG Temuco - Chile

A validation assay, based on the golden standard test, was designed with the objective of estimating parameters and comparing the performance of serology tests for the diagnosis of bovine brucellosis. Six tests were assayed: two competitive ELISA tests, three indirect ELISA tests and a Fluorescence Polarization Assay. The study followed the recommendations of the OIE.

An infected bovine reference population was confirmed by isolation of *Brucella abortus* from 426 sera samples. In addition, the 40 Rose Bengal and Complement Fixation tests that were performed yielded positive results. The sera samples were collected from 11 infected herds located in regions10 and 11 and the Metropolitan region. The brucellosis free reference population was confirmed by 1,126 bovine sera samples from 12 officially brucellosis free herds, collected in regions 9 and 10 which tested negative using official tests. The tests were carried out according to the protocols recommended by the manufacturers.

The following diagnostic parameters were estimated: diagnostic sensitivity and specificity, Jouden Index and ROC curve. The parameters were initially estimated according to the cut-off points suggested by the manufacturer and later optimized using Minimum Misclassification Costs. The estimations were calculated with Stata version 8.2 for Windows, CMDT version 1.0 and Epidat version 3.1.

The estimated parameters using the optimization of the cut-off points were significantly better than the estimates based on those suggested by the manufacturers. Diagnostic sensitivity ranged from 96.1% (94, 3-97, 8) to 98.5% (97, 3-

99, 7) which was not significant. On the other hand, diagnostic specificity ranged from 96.4% (95, 2-97, 5) to 99.4% (98, 9-99, 8) which was significant ($p<0.05$). The Jouden Index ranged from 0.94 (0, 92-0, 96) to 0.98 (0, 97-0, 99) and showed significant differences ($p<0.05$). Significant differences were also observed when comparing ROC curves of the tests. The competitive ELISA tests showed better performance compared to the indirect ELISA tests. The performance of the Fluorescence Polarization Assay did not show significant differences compare to ELISA tests. The observed estimations should be weighed against a possible selection bias inherent to this model of validation study.

Lombard M, Dodet B, the OIE Scientific & Technical Department (eds): First International Conference of the OIE Reference Laboratories and Collaborating Centres. Dev Biol (Basel). Basel, Karger, 2007, vol 128, p 173.

Infectious Disease Diagnosis in Cultured Aquatic Organisms, More than a Single Technique

J. Cáceres Martínez[1,2], R. Vásquez Yeomans[2], A.M. García Ortega[2,3]

[1] Laboratorio de Biología y Patología de Organismos Acuáticos del Departamento de Acuicultura del CICESE. Laboratorio de Referencia de enfermedades de Moluscos PRONALSA-CICESE, Ensenada, B. C., México 22860 - [2] Instituto de Sanidad Acuícola, México 22800 - [3] Centro Regional de Estudios y Diagnóstico Fitosanitario del Comité Estatal de Sanidad Vegetal de Baja California, San Felipe Mexicali, B. C., México

The discovery of *Thermus aquaticus* in 1969 opened the door to one of the most important biotechnological advances of the 20th century. This bacterium is a source of the thermo-stable DNA enzyme Taq polymerase which allowed for the development of the Polymerase Chain Reaction (PCR) to be used to obtain DNA copies from a single DNA fragment. Application of this technique to the identification of pathogens is widely recognized and has a direct impact on the precision of infectious disease diagnosis around the world. Unfortunately, an overvaluation of this technique often leads to discrimination against the conventional diagnostic techniques of classical bacteriology, parasitology and histopathology. This results in important diagnostic mistakes and economic losses. We analysed different technical aspects related to the application of PCR for the diagnosis of infectious diseases in aquatic organisms that may result in false positives or false negatives. Moreover, we looked at the importance of the sanitary history of a sampled population, sampling criteria, the use of conventional diagnostic techniques and broad pathology criteria in making a diagnosis. PCR is a powerful technique for the diagnosis of infectious diseases in aquatic organisms and several of its inherent technical sources of error will be eliminated. However, it is just one technique of many available to pathologists. Results from different techniques and the use of broad pathology criteria are essential when formulating a dependable diagnosis.

Lombard M, Dodet B, the OIE Scientific & Technical Department (eds): First International Conference of the OIE Reference Laboratories and Collaborating Centres. Dev Biol (Basel). Basel, Karger, 2007, vol 128, p 174.

Mapping of a SiRNA-Susceptible Locus on Morbillivirus Nucleoprotein mRNA

D. Keita, R.S. Almeida, E. Albina, G. Libeau

CIRAD Département-EMVT, UPR15, Montpellier, France

Morbilliviruses are pathogens with a major socio-economical impact for humans and animals. Peste des petits ruminants (PPR) is a severe disease of small ruminants, while rinderpest (RP) affects even-toed ungulates, most commonly cattle and buffalo. Both diseases have high morbidity and mortality rates. Measles (Me) is an acute disease of humans that claims about half a million lives annually, principally among young children of the less developed countries. Efficient vaccines are available for the three diseases. However, common characteristics of morbillivirus infections include being highly infectious and the absence of an effective and specific treatment. We therefore considered that the strategy of RNA interference could be used to control morbillivirus infections. We assumed that siRNA against viral nucleoprotein (N) mRNA could induce a strong inhibitory effect since morbillivirus N plays a central role in viral transcription and replication. We developed cellular and molecular approaches in order to identify small interfering RNA-susceptible loci on the morbillivirus N gene. RNA interference is a natural cell mechanism by which small 20 nucleotide siRNA can downregulate the expression of a gene by targeting its mRNA. We thus identified synthetic siRNA effective for the three morbilliviruses PPRV, RPV and MeV.

Lombard M, Dodet B, the OIE Scientific & Technical Department (eds): First International Conference of the OIE Reference Laboratories and Collaborating Centres. Dev Biol (Basel). Basel, Karger, 2007, vol 128, p 175.

The OIE Reference Laboratory for Trichinellosis: Fourteen Years of Activity

E. Pozio

Istituto Superiore Di Sanità, Rome, Italy

More than 180 researchers from 56 countries have relied on the OIE Reference Laboratory for Trichinellosis (OIE-RLT) to identify isolates, to receive reference strains, antigens, DNA, PCR primers, epidemiological information, and sera from infected animals and humans since 1992. The OIE-RLT database can be consulted by accessing the web-site www.iss.it/site/Trichinella/index.asp. The OIE-RLT is also the official reference laboratory of the International Commission on Trichinellosis and it acts as the Community Reference Laboratory for Parasites of the EU. To date, more than 1,700 *Trichinella* isolates of human and animal origin from throughout the world have been examined and identified by new procedures developed at the OIE-RLT. More than 80 *Trichinella* isolates are maintained in vivo. These data have been vital to the establishment of a new taxonomic system for *Trichinella*, which has had a great impact on the epidemiology and management of these parasites infecting mammals, birds and reptiles. The identification of parasites as domestic or sylvatic types has helped veterinarians to plan control programmes, to select target animals, and to adopt the most useful techniques for the veterinary control of meat safety. Serological investigations on human and animal sera carried out with homemade and commercial tests have revealed problems with using serology for the diagnosis of this infection in humans and animals due to long window periods before the appearance of detectable circulating antibodies and the fast disappearance of circulating antibodies in of spite the presence of infecting larvae in the muscles.

Lombard M, Dodet B, the OIE Scientific & Technical Department (eds): First International Conference of the OIE Reference Laboratories and Collaborating Centres. Dev Biol (Basel). Basel, Karger, 2007, vol 128, p 176.

Fine Differentiation of *Mycoplasma Mycoides* Subsp. *Mycoides* Sc Strains by Multilocus Sequence Typing

A. Yaya[1], L. Manso-Silvan[2], A. Peyraud[2], F. Thiaucourt[2]

[1] LANAVET Garoua Cameroon - [2] CIRAD UPR15 Montpellier France

Contagious bovine pleuropneumonia (CBPP) is a disease affecting cattle and domestic buffaloes caused by a mycoplasma: *Mycoplasa mycoides* subsp. *mycoides* SC (MmmSC). The development of new tools for typing MmmSC strains, capable of tracing back the origin of the outbreaks, is much needed. An initial multilocus sequence typing (MLST) scheme was designed in 2003 in our laboratory (Lorenzon et al. 2003). Since then, the complete sequence of reference strain Pg1 has been published and the whole genome of a pathogenic strain of African origin is under assembly in our laboratory. The comparison between these two sequences allowed for the selection of new loci for MLST typing of MmmSC strains. Thirty potential loci were chosen on housekeeping genes, on genes coding for surface proteins and on non-coding sequences. Eight were selected and the analysis resulted in an unprecedented discriminative power to differentiate and catalogue MmmSC strains. The 48 strains studied could be differentiated into 32 MLST profiles. Strains isolated recently in Europe belonged to a specific group, suggesting that recent CBPP outbreaks represented a resurgence of the disease and were not due to an introduction. Some correlations were also found between older European strains and strains found in Southern Africa which is in accordance with the introduction of the disease in this region during the 19th century. All strains isolated recently in Eastern Africa belonged to the same MLST type, an indication of a clonal expansion in agreement with the recent re-introduction of the disease.

Lombard M, Dodet B, the OIE Scientific & Technical Department (eds): First International Conference of the OIE Reference Laboratories and Collaborating Centres. Dev Biol (Basel). Basel, Karger, 2007, vol 128, p 177.

Evaluation of a Competitive ELISA for Detection of Vesicular Stomatitis Virus Antibodies in Horses during the 2004 and 2005 Outbreak in the United States of America

T.L. Buffington, B.L. Kesl, M.L. Redlinger, T.J. Beach, S.L. Swenson, B.J. Schmitt

National Veterinary Services Laboratories, Veterinary Services, Animal and Plant Health Inspection Service, US Department of Agriculture, Ames, IA, United States

Vesicular Stomatitis (VS) is a vesicular disease of horses, pigs, cattle, and other ruminants that is endemic in parts of North, Central and South America. It is imperative to have a rapid diagnosis to avoid costly quarantines and trade restrictions when VS is suspected. The competitive enzyme-linked immunosorbent assay (CELISA) was developed for use as a screening assay for detection of antibodies against vesicular stomatitis New Jersey and Indiana-1 virus serotypes. The development of this assay has provided an alternative to labor intensive serum neutralization and complement fixation (CF) tests for movement testing. The CELISA also supplements the CF test as a sensitive and specific tool for providing same day diagnosis for animals suspected of having VS.

During the 2004 and 2005 outbreaks of VS in the United States, the CELISA was used by the NVSL as a screening tool for detection of VS antibodies in clinically affected horses at the National Veterinary Services Laboratories. The CF test was also performed on the same samples gathered during investigations on suspected vesicular stomatitis cases. The data accumulated for both the CELISA and CF tests illustrate the value of the CELISA test in the early detection of positive animals.

List of Participants

SPEAKERS

BELAK Sandor, R&D Division, National Veterinary Institute, Ulls väg 2 B, SE-751 89 Uppsala, Sweden
BRANCO DE FREITAS MAIA Lucia Maria, Ministério da Agricultura, Pecuária e Abasteciment, Esplanada dos Ministérios Bl D, Anexo B Sala 432, Brasilia, DF 70043-900, Brasil

CAMUS Emmanuel, CIRAD-EMVT, Campus International de Baillarguet, TA 30/B, FR-34398 Montpellier Cedex, France
CAPORALE Vincenzo, Istituto Zooprofilatico Sperimentale, 1 Via Campo Boario, IT-64100 Teramo, Italy
COUSINS Debby, Application and Linkage, Australian Biosecurity CRC, GPO Box U1987, 6152 Bentley, Perth, WA 6845, Australia

DEHAUMONT Patrick, AFSSA-LERMVD, BP 90203, FR-35302 Fougères Cedex, France
DREW Trevor, VLA Weybridge, Woodham Lane, New Haw, Addlestone, Surrey KT15 3NB, U.K.

EDWARDS Steven, VLA Weybridge, New Haw, Addlestone, Surrey KT15 3NB, U.K.

GARIN-BASTUJI Bruno, AFSSA, 23 av. du General de Gaulle, FR-94706 Maisons-Alfort Cedex, France

HILL Barry, Cefas, Barrack Road, The Nothe, Weymouth, Dorset DT4 8UB, U.K.

JEGGO Martyn, CSIRO, Fort Lignston, Geelong, VIC 3030, Australia

LUBROTH Juan, Animal Health Service, FAO, Viale Delle Terme Di Casacalla, IT-00161 Roma, Italy

METTENLEITER Thomas, Friedrich-Loeffler Institute, Boddenblick 5A, DE-17493 Greifswald, Insel Riems, Germany

NICOLA Ana María, Dilab-Senasa, Av. Fleming 1653, 1640 Martinez, Buenos Aires, Argentina.

PATON David, Institute for Animal Health, Pirbright Laboratory, Ash Road, Woking, Surrey GU24 0NF, U.K.
PEARSON James, OIE, 4016 Phoenix Street, Ames, Iowa 50014, USA

SCHMITT Beverly, Diagnostic Virology Lab., National Veterinarian Services Laboratories, 1800 Dayton Ave, Ames, Iowa 50201, USA
SPIESER Jean-Marc, Department for Biological Standards and OMCL Network, European Directorate for the Quality of Medicine, B.P. 907, FR-67029 Strasbourg Cedex 1, France

UNGER Hermann, International Atomic Energy Agency, Wagramer Strasse 5, PO Box 100, AU-1400 Vienna, Austria

ZEPEDA Cristobal, International Activities, USDA/APHIS, 2150 Centre Av. Bldg B, Fort Collins, CO 80526, USA

EXPERTS

ARZUL Isabelle, Ifremer, Ronce les Bains, FR-17390 La Tremblade, France

BANAI Menachem, Department of Bacteriology, Kimron Veterinary Inst., PO Box 12, IL-50250 Beit Dagan, Israel
BANKS Malcolm, VLA Weybridge, Veterinary Laboratories Agency, New Haw, Addlestone, Surrey KT15 3NB, U.K.
BARRAT Jacques, Epidemiology and Wildlife Diseases Unit, AFSSA-Lerpas, Zoopole Agricole et Veterinaire, BP 40009, FR-54690 Paris, France
BEIER Dagmar, Friedrich-Loeffter Institute, Seestr. 55, DE-16868 Wusterhausen, Germany
BERGMANN Ingrid, Centro Panamericano de Febre Aftosa OPS/OMS, Prudente de Morais 331/102, Rio de Janeiro, RJ 22420-041, Brasil
BERNARDELLI Amelia, Sevicio Nacional de Sanidad y Calidad Agroalimentaria, Senasa, Avenida Alecsander Fleming, 1653 Martinez, AR-1640 Buenos Aires, Argentina
BÖLSKE Göran, SVA, SE-75189 Uppsala, Sweden

CAPUCCI Lorenzo, Istituto Zooprofilattico, Biancmi 9, IT-25124 Brescia, Italy
CLAES Filip, Institute of Tropical Medicine Antwerp, Nationale Straat 155, BE-2000 Antwerpen, Belgium
CZUB Stefanie, BSE Reference Laboratory, Canadian Food Inspection Agency, 9-1 Township Road, Lethbridge, Alberta T1J 3Z4, Canada

DAVIES Robert, VLA Weybridge, Woodham Lane, New Haw, Addlestone, Surrey KT15 3NB, U.K.

ELLIS William, Stoney Road, Stormont, Belfast BT4 3SD, Northern Ireland, U.K.

FADLY Aly, USDA-ARS Avian Disease and Oncology Lab., 3606 East Mount Hope Road, East Lansing, MI 48823, USA
FAIRBROTHER John, Université de Montréal, 3200 Sicotte, St-Hyacinthe, Québec J25 2M2, Canada
FOOKS Anthony, Rabies and Wildlife Zoonoses, Veterinary Laboratories Agency, New Haw, Addlestone, Surrey KT15 3NB, U.K.

GAJADHAR Alvin, Centre for Food-Borne & Animal Parasitology Canadian Food Inspection Agency, 116 Veterinary Rood, Saskatoon, Saskatchewan SK S7N 2R3, Canada
GERDES Truuske G.H., Ondersteroort Veterinary Institute, Old Soutpans Road, 110 Onderstepoort Gauteng- Pretoria, South Africa

HARDER Timm, Friedrich-Loeffler Institute, Boddenblick 5A, Insel Riemsd, DE-17493 Wusterhausen, Germany
HARTKEERL Rudolf A., Leptospirosis Reference Lab., Royal Tropical Institute, Meiberg Dreef 32, NL-1105 AD Amsterdan, Netherlands
HEATH Peter, VLA Weybridge, Rougham Hill, Bury St-Edmunds, Suffolk IP33 2RX, U.K.
HILL Richard, Center for Veterinary Biologics, 510 S., 17th Street, Ste 104, Ames, Iowa 50010, USA

JAHANS Keith, VLA Weybridge, Woodham Lane, New Haw, Addlestone, Surrey KT15 3NB, U.K.

KAMIYA Masao, Food Agriculture of the UM, 582 Midorimachi, Bunkyodai, Ebetsu-shi, Hokkaido 069-8501, Japan
KANCHANAKHAN Somkiat, AAHRI, Paholionthin Rd, Jatuchak, Bangkok 10900, Thailand
KIBENGE Frederick, Atlantic Veterinary College, University of Prince Edward Island, Dept of Pathology and Microbiology, 550 University Ave., Charlottetown, Prince Edward Island C1A 4P3, Canada

KONDO Takashi D, Epizootic Research Center, Equine Research Institute, Japan Racing Association, 1400-4, Shiba, Shimotsuke, Tochigi 329-0412, Japan

KRAMPS Johannes, CIDC, Lelystad, Houtribweq St. 39, PO Box 2004, NL-8203 AA Lelystad, The Netherlands

LIBEAU Genevieve, Lab RP/PPR, CIRAD/EMVT, TA30B, Campus Intern. de Baillarguet, FR-34398 Montpellier Cedex, France
LIGHTNER Donald, The University of Arizona, 1117, East Lowell St., Tucson, AZ 85721, USA
LUBISI Baratang, Ondestepoort Veterinary Institute, Private Bag X05, Onderstepoort 110, South Africa

MELLOR Philip, Institute for Animal Health, Pirbright Laboratory, Ash Road, Woking, Surrey GU24 0NF, U.K.
MELZER Falk, NRL Brucellosis Germany, OIE Collaborating Centre, Friedrich-Loeffler Institute, Naumburger Str. 96a, DE-7743 Jena Thuringia, Germany
MOENNIG Volker, Institute of Virology University of Veterinary Medicine, Buenteweg 17, DE-30559 Hannover, Germany
MUCKLE Anne, OIE Reference Lab. for Salmonellosis, Public Agency of Canada, Guelph, Ontario N1G 3W4, Canada
MUMFORD Jennifer, Dept. Veterinary Medicine Cambridge University, Madingley Rd, Cambridge CB3 0ES, UK
MURAKAMI Kenji, National Institute of Animal Health, 3-1-5 Kannodai, Tsukuba, Ibaraki 305-0856, Japan

NIELSEN Klaus, Canadian Food Inspection Agency, 3851 Fallowfield Rd, Nepean, Ontario K2H 8P9, Canada

OLESEN Niels, Danish Institute, Hangovej 2, DK-8200 Arrhus, Denmark
OSTLUND Eileen, Equine and Ovine Viruses Section, Diagnostic Virology Laboratory, 1800 Dayton Road, P.O. Box 844, Ames, Iowa 50010, USA
OURA Chris, Epidemiology and Head of Non-Vesicular Reference Labs, Institute for Animal Health, Ash Rd, Woking, Surrey GU22 8AY, U.K.

PANIGRAHY Brundaban, NVSL APHIS-USDA, 3603 Annear Strret, Ames, Iowa 50014, USA
PANIN Alexander, VGNKI, Zvenigorodskoesh. 5, RU-123022 Moscow, Russia
PASICK John, Classical Swine Fever & Avian Disease Unit, 1015 Arlington St., Winnipeg, Manitoba R3E 3M4, Canada
PAVLIK Ivo, Mycobacteriology, Veterinary Reseach Institute, Hudcova 70, CZ-60200 Brno, Czech Rep.
PEJSAK Zygmunt, Swine Diseases Department, National Veterinary Research Institute, Partyzantów 57, Pulawy, POL-24-100 Lublin, Poland
PERRIN Gérard, Agence Française Securité Sanitaire des Aliments, Pied de Fard, FR-79012-Nior Cedex, France
POMBO BOTELHO Ana Rosa, Bacteriology Dept, Laboratório Nacional Investigação Veterinária LNIV, Estrada de Benfica, 701, PT-1549 011 Lisboa, Portugal
POZIO Edoardo, OIE Reference Laboratory for Trichinellosis, Istituto Superiore di Sanitá, Viale Regina Elena 299, IT-00161 Roma, Italy

RENSTROM Lena, SVA, National Veterinary Institute, SE-751 89 Uppsala, Sweden
ROEST Hendrik, CIDC Lelystad, Edelhertweg 15, NL-8219 PH Lelystad, Netherlands
ROSSELA Lelli, National Reference Centre Exotic Diseases of Animals IZS A.M., Istituto Zooprofilattico Sperimentale Abruzzo e Molise "G. Caporale", Campo Boario, IT-64100 Teramo, Italy

SABETA Claude, Onderstepoort Veterinary Institute, (OVI), Virology Dept, Old Soutpan Rd, P.B. X05, Onderstepoort 0110, Pretoria, South Africa
SAIF Linda, FAHRP, 1680 Nadisen Ave, Wooster, Ohio 44691, USA
SAIF Yehia M., The Ohio State University, Fool Animal Heath Research Program, OARDC, Wooster, OH 44691, USA
SAMARTINO Luis, Instituto Nacional de Tecnologia Agropecuaria, INTA, Las Caranãs y los Reseros, Boron, 1712 Buenos Aires, Argentina

SAVINI Giovanni, Virology Dept. OIE, Instituto Zooprafilattico "Abruzzo e Rousé", Via Campo Boario, IT-64100 Teramo, Italy
SCACCHIA Massimo, Fiscal Federal Agropecuario, Reference Labs, Campo Boario, IT-64100 Teramo, Italy
STACK Judy, Veterinary Laboratories Agency, New Haw, Addelestone, Surrey KT15 3NB, U.K.

TEKO-AGBO Assiongbon, Inter States of School of Veterinary Siences and Medicine, EISMV, PO Box 5077, Dakar, Senegal
THIAUCOURT François, Contagious Bovine Pleuropneumonia Reference Center, CIRAD, Campus Baillarguet, TA 30/B, FR-34398 Montpellier, France
TIMONEY Peter, University of Kentucky, Gluck Equine Research Ctr., Lexington, Kentucky 40546, USA

VARSHOVI Hamid Reza, Razi Institute, Hessarak, PO Box 319 75148 Teheran, Iran

YAMADA Shunji, National Institute of Animal Health, 6-20-1 Josui-honcho, 187-0022 Tokyo, Japan
YOKOYAMA Takashi, Prion Disease Research, National Institute of Animal Health, 3-1-5 Kannodai, 305 0856 Tsukuba, Ibaraki, Japan

ORGANISATIONS

ARRUA Victor, IICA, Bernardo de Irigoyen, 88, Maidana, Buenos Aires AR, Argentina

GENOVESE LINARES Miguel Angel, OPS, Av. Pres Kenedy, 7778, Florianopolis, SC 25040-004, Brasil

LOMBARD Michel, IABs (International Association for Biologicals), 22, rue Crillon, FR-69006 Lyon, France

WASHINGTON LÓPEZ Jorge, Centro Pan-Americano de Febre Aftosa (OPAS/OMS), Rua Souza Lima, 51, apt. 1001, Copacabana, Rio de Janeiro, RJ 22081-0010, Brasil

OIE

BARCOS Luis, OIE, Cerviño 3101, Buenos Aires, 1425 Argentina
BEN-JEBARA Karim, Departamento Científico OIE, 12, rue de Prony, FR-75017 Paris, France
BORGEAUD Ines, Cerviño 3101, 1425 Buenos Aires, Argentina
BRÜCKNER Gideon, Scientific and Technical Department, OIE, 12, rue de Prony, FR-75017 Paris, France

ERLACHER-VINDEL Elisabeth, Scientific and Technical Department, OIE, 12, rue de Prony, FR-75017 Paris, France

LINNANE Sara, OIE, 12 rue de Prony, FR-75017 Paris, France

MTEI Bonaventure, Regional Representative Botswana OIE, Orapa Close Plot H 8850, Gaborone, Botswana
MURRAY Gardner J., P.O. Box 4215, Western Creek, 2611 Australia

OREAMUNO Jose Joaquin, OIE, Calle La Petra, San Rafael de Oreamuno, 7050 Cartago, Costa Rica

PALMAS Alicia Susana, Cerviño 3101, 1425 Buenos Aires, Argentina
PLANTE Caroline, OIE, 12, rue de Prony, FR-75017 Paris, France

SODJI Kokoé, OIE, 12 rue de Prony, FR-75017 Paris, France

VALLAT Bernard, Director General, OIE, 12, rue de Prony, FR-75017 Paris, France

PARTICIPANTS

ACOSTA Ladislao, Pathology and Food Control Laboratory, Servicio Nacional de Calidad y Salud Animal, Ruta Mcal Estigarribia Km 10,5, Não San Lorenzo, Paraguay
ALBIERI CAMILLO Soraya Cecilia, Mapa, Rua Raul Ferrari S/N, Campinas, SP 13100-105, Brasil
ALEKSEY Shcherbakov, FGI Arriah, Yur'evets, RU-600901 Vladimir, Russia
ALVES DE SOUZA Marcilia Maria, Adagro/PE, Rua Frei Caneca 429 Centro, São Lourenç Da Mata, PE 54735-796, Brasil
ALVES MACIEL Gabriel, Mapa, Ministério Da Agricultura, Esplanada Dos Ministérios, Bloco D Anexo B Sala 406, Brasília, DF 70043-900, Brasil
ALVES MARTORELLI Luzia Dátima, CCZ/SP, Rua Santa Eulália, 86-Santana, São Paulo, SP 02031-020, Brasil
AMORIM SCHIAVO Paula, Mapa, Av. Almirante Barroso 1234, Belem, PA 66095-000 Brasil
ANDORFFY SARAIVA Carlos Pedro, Bayer S.A., Rua Edú Chaves, Porto Alegre, RS 360-9024-062, Brasil
ANDRADE JÚNIOR Antônio Araújo, Mapa, Esplanada dos Ministérios, Ed. Sede, Bl-D, Anexo A, Sala 448-A, Brasilia, DF 70.043-900, Brasil
ARAÚJO COSTA Héber Brenner, Ministério da Agricultura, Pecuária e Abastecimento, Esplanada dos Ministérios, 4º Andar- Ala B Sala 448-B, Brasília, DF 70 043-900, Brasil
ARRUDA DE CARVALHO Augusto Vinícus, Lanagro/MG, Av. Rômulo Joviano s/n Faz. Modelo, Cx Postal 50, Pedro Leopoldo, MG 33600-000, Brasil
ARTIAGA Maria Soares, Ministério da Agricultura, Pecuária e Abasteciment, Sqn 311 Bloco L, Aptº 103, Asa Norte, Brasília, DF 70000-000, Brasil
AUBERT Michael, Laboratorio, Rte des Chappes, Sophia Antipolis, FR-06000 Nice, France
AYLING Roger, VLA Weybridge, Woodham Lane, New Haw, Addlestone, Surrey KT15 3NB, U.K.

BACK Alberto, MercoLab, Rua São Paulo 645 Apt. 201, Cascavel, PR 85 801-020, Brasil
BAGOLIN PALMEIRA Andrá Luiz, Ministério da Agricultura, Rua Curitiba 1275 D, Ap 104 Bloco A, Chapecó, SC 89812-150, Brasil
BALIZA PEREIRA Heloiza H., Vallée S.A., Av. Hum 1500, Distrito Industrial, Montes Claros, MG 39404-003, Brasil
BARBOSA PINHEIRO Diana Sione, Lanagro/PE, Rua Dom Manoel de Medeiros S/N, Bairro Dois Irmãos, Recife, PE 52171-030, Brasil
BARBOZA TAVARES Luiz Carlos, Defesa Agropecuarea e Florestal do Espiri Santo, Rua Minas Gerais Nº 660 Bloco1, Aptº 402, Laranjeiras II, Serra ES 29165-725, Brasil
BARRETO DA COSTA ORTIZ Silvia Colleta, Cobea, Rua d. Mariquita Julião 146, Murumbi, São Paulo, SP 05656-070, Brasil
BARROS DE FREITAS Josinete, Mapa, Sqs 410 Bloco S Apto 202, Asa Sul, Brasilia, DF 70290-196, Brasil
BASHIRUDDIN John B., Institute for Animal Health, Ash Road, Pirbright, Woking, Surrey GU24 0NF, U.K.
BECKER João M., Lanagro/RS, Anita Garibaldi, 2120 / 1315, AP 1315, Porto Alegre, RS 90480-200, Brasil
BELÉM Andréa, UFAM, Rau Luis Antony 842 Centro, Manaus, Manaus, AM 69010-100, Brasil
BELLEMAIN Veronique, Ecole Nationale des Services Vétérinaires, 1, av. Bourgelat, FR-69280 Marcy-L´Etoile, France
BIESDORF DORNELES RODRIGUES Sonia Maria, Centro de Diagnóstico Marco Enriett, Rua Jalme Balão 575, Campus 1-Ufpr, Juvevê, Curitiba, PR 80040-340 Brasil
BLOCH JUNIOR Carlos, Embrapa, Recursos Genéticos, Parque Estação Biológica W5, Norte Final, 70770-900, Brasil
BONNET ALVARENGA Marcelo, Engenheiro de Alimentos, Mapa, Condomínio Village de Alvorada I, Casa 8 Lago Sul, Brasilia, DF 70000-000, Brasil

BORISOV Vladimir, FGI Arriah, Yur'Evets, RU-600901 Vladimir, Russia
BOTREL COUTINHO Marcio Danilo, Rodovia Br 365 Km 615, Bairro Conj. Alvorada, Uberlandia, MG 38407-180, Brasil
BRENTANO Liana, Pesquisadora III, Embrapa, Rua Raul Ferrari S/N, Campinas, SP 13100-105, Brasil
BURMEISTER MARTINS Valéria, Ministério da Agricultura, Pecuária e Abasteciment, Ville Blanche III Bloco B Apt° 1, Parque Esplanada III, Valparaíso De Goiás, GO 72876-341, Brasil

CACERES MARTINEZ Jorge, Laboratorio de Referencia de Enfermedades de Moluscos, Pronalsa Mexico, Km 107 Carret, Tijuana, Ensenada, Baja–California, Mexico
CAETANO JUNIOR Jorge, Das, Mapa, Shis QI 25, Conj. 11, Casa 01, Brasília, DF 71.660-310, Brasil
CALDEIRA George A. Vitor, Lanagro/MG, Av. Rômulo Joviano S/N, Faz. Modelo, Cx Postal 50, Pedro Leopoldo, MG 33600-000, Brasil
CAMPANA FURLANETTO Sônia Maria, Unesp, Rua Luiz Fernando Rodrigues, 1701, Vila Boa Vista, Campinas, SP 13174-532, Brasil
CARVALHO DE OLIVEIRA Rodnei, Cidasc, Arnaldo Bonckewtz, 44, Casa, São José, SC 86.103-215, Brasil
CASALI FUENTES Marcelo, Servicio Nacional de Pesca, Victoria 2832, Valparaíso, 234-0159 Valparaíso província de Valparaíso, Chile
CASSARO VILLALOBOS Eiana M., Instituto Biológico, Av. Conselheiro Rodrigues Alves, 1252, São Paulo, SP 04014-002, Brasil
CASTELLO BRANCO PERMAN Raiolanda, Lanagro/PE, Rua Dom Manoel de Medeiros S/N, Recife, PE 52171-030, Brasil
CAVALCANTE DE CARVALHO Rosângela, Adapi/PI, Rua Chico Doca n° 641, Bairro Acarate, Teresina, PI 64003-830, Brasil
CAVANI Ricardo, Seara Alimentos, Rua Guilherme Tizzioti 716, Sales Oliveira, SP 14660-000, Brasil
CHO Donghee, National Veterinary Research & Quarantine Service, 480 Anyang 6-dong, Anyang-si, Kyounggi-do, 430-824 Anyang, Korea
CIACCI ZANELLA Janice Reis, Embrapa, Br 153, Km 110, Distrito de Tamanduá, Cx Postal 21, Concordia, SC 89700-000, Brasil
COELHO MOTA Pedro Moacyr P., Ministério da Agricultura, Av. Rômulo Joviano s/n Faz. Modelo, Cx Postal 50, Pedro Leopoldo, MG 33600-000, Brasil
CORBELLINI Luís G., Universidade Federal do Rio Grande do Sul (UFRGS), Av. Bento Gonçalves 9090, Porto Alegre, RS 90660-280, Brasil
COSTA GUEDES Sebastião, GIEFA-Grupo Inter-Americano para Erradicação da Fe, Rua do Rócio 313, cj 92 Vila Olímpia, São Paulo, SP 04552-000, Brasil
COUBE BOGADO Sergio, VLA Weybridge, Rua Mariz e Barros, 152- Apto. 201, Niteroi, RJ 24220-121, Brasil
CUNHA CALLADO Ana Karina, Fiscal Federal Agropecuário, Lanagro/Pe, Rua Dom Manoel De Medeiros S/N, Bairro Dois Irmãos, Recife PE 52171-030, Brasil

DA SILVA JÚNIOR Anísio Antônio, Vetbrands Animal Health, Campinas, Campinas, SP 13010-110, Brasil
DE ALMEIDA ROSA Julio Cesar, PVDF, Estrada Municipal do Conde 6000, Eldorado do Sul, RS 92990-000, Brasil
DE ARAUJO CALDAS Aluisio Fanklin, Mapa - Sqn 112 Bloco G Apto 601, Asa Sul, Brasilia, DF 70762-070, Brasil
DE ASSIS SANTANA Vania Lucia, Mapa/Lanagro/Pe, Rua Professor Francisco Pessoa De Melo N° 851, Candeias, Jaboatão, PE 54450-180 Brasil
DE AZAMBUJA Germano Augusto, SFA/SP, Rua Londrina, 410 Apt° 61, Rudge Ramos, São Bernardo Do Campo, SP 09635-100, Brasil
DE AZEVEDO RUIZ Vera Letticie, Instituto Biológico, Avenida Conselheiro Rodrigues Alves, 1252, Laboratório de Doenças de Suínos, São Paulo SP 4014-002 - Brasil
DE CARVALHO FILHO Maurício Baltazar, Lanagro/MG, Av. Rômulo Joviano s/n Faz. Modelo, Cx Postal 50, Pedro Leopoldo, MG 33600-000, Brasil
DE CARVALHO MARQUES SERQUEIRA Marco A., Mapa, Esplanada Dos Ministerio Bloco D Anexo, sala 432, Brasilia, DF 70043-900, Brasil
DE CASTRO CHAVES Silvio Orlan, Mapa, Av. Almirante Barroso, 1234, Belem, PA 66095-000, Brasil

DE CASTRO OLIVEIRA Ricardo, Vallée SA, Av Hum, 1500, Distrito Industrial, Montes Claros, MG 39404-003, Brasil
DE CENA OLIVEIRA Marcia Cristina, Embrapa/CPPSE, Rodovia Washinton Luiz, Km 234, São Carlos, SP 13560-970, Brasil
DE DARVALHO BRAGA Alexandre, IPVDF, Estrada Municipal Do Conde 6000, Eldorado Do Sul, RS 92990-000, Brasil
DE FARIAS CORDEIRO Claudia, Lanagro/PE, Rua Dom Manoel de Medeiros S/N, Recife, PE 52171-030, Brasil
DE LIMA MIGNOT João P., Esterilix, Av. Brigadeiro Luis Antonio, 613, 90 Andar, São Paulo, SP 1317-000, Brasil
DE MELO Carlos Alberto, Mapa, r. Bocaiuva 2245, Apt 401Florianopolis, SC 88015-510, Brasil
DE OLIVEIRA MENDONÇA Andre, Mapa, Rua Raul Ferrari, S/N, Campinas, SP 13100-105, Brasil
DE SIQUEIRA ALENCAR Cid Aristóteles, Mapa/Lanagro/PE, Rua Dom Manoel De Medeiros S/N, Bairro Dois Irmãos, Recife, PE 52171-030, Brasil
DE SOUZA MORAES Hamilton L., UFRGS, Gen. Lima e Silva 130, Apto. 68, Cidade Baixa, Porto Alegre, RS 91000-000, Brasil
DE STEFANO Eliana, Instituto Biológico, Rua Botupuca, 194, São Paulo, SP 1536-010, Brasil
DEJESUS FERREIRA Ângela Maria, Lanagro/MG, Av. Rômulo Joviano s/n Faz. Modelo- Cx Postal 50, Pedro Leopoldo, MG 33600-000, Brasil
DEON KICH Jalusa, Centro Nacional de Pesquisa de Suino e Aves, Distrito de Tamanduá, Km 110 BR 153, Cx Postal 21, Concordia, SC 89700-000, Brasil
DESTRA Karen C., Provet, Travessa 14 de Abril, 1235, Belem, PA 66035-000, Brasil
DI PILLO Fulvia, Mapa, Rua Raul Ferrari S/N, Campinas, SP 13100-105, Brasil
DIAMANTINO FEIJO Leandro, Ministério da Agricultura, Pecuária e Abasteciment, Esplanada Dos Ministérios, 4º Andar, Ala B Sala 448 B, Brasília, DF 70 043 900, Brasil
DOS SANTOS Fernando L., UFRPE, Rua Paes Cabral 617, Recife, PE 50630-170, Brasil
DRIEMEIER David, UFRGS, Av. Bento gonçalves 9090, Bairro Agronomiia, Porto Alegre, RS 91540-000, Brasil
DURIGON Edison L., Univ. São Paulo, Av. Prof. Lineu Prestes 1374, São Paulo, SP 550897, Brasil

EKGATAT Monaya, Dpt of Livestock Development, National Institute of Animal Health, Kasetklang Bangkhen, Tatuchak, Bangkok 109000, Thailand
EVANS Brian R., Canadian Food Inspection Agency, Room 146 W, 59 Camelot Drive, Ottawa, Ont. K1A 0Y9, Canada

FAGUNDES NASCIMENTO Kelly, Mapa, Benjamin Anahas, 255 B. Paraíso, Belo Horizonte, MG 30.270-130, Brasil
FAVORETTO LAZARINI Silvana Regina, USP/SP, Av. Prof. Lineu Prestes, 1.374, Cidade Universitária, São Paulo, SP 05508-900, Brasil
FAYER CALEGARIO Fagoni, Embrapa-Meio Ambiente, Rua Latino Coelho Nº 1301, Bloco B Aptº 9, Campinas, SP 13087-010, Brasil
FEDOU Jean-Claude, Ceva Saúde Animal Ltda, Rua Santa Justina 660, Vila Olímpia, São Paulo, SP 04545-042, Brasil
FELTRIN DAMBRÓS Régia Maria, Cidasc/Cedisa, Br 153 Km 110, Vila Tamanduá, Concórdia, SC 89700-000, Brasil
FERNANDES CAMARGOS Marcelo, Lanagro/MG, Av. Rômulo Joviano s/n Faz. Modelo, Cx Postal 50, Pedro Leopoldo, MG 33600-000, Brasil
FERNANDES DE ALMEIDA Marilene, CCZ-SP Prefeitura, Rua Sta. Eulália 86, Santana, São Paulo, SP 02031-020, Brasil
FERREIRA DE FIGUEIREDO Vera Cecília, Ministério da Agricultura, Pecuária e Abasteciment, Esplanada Dos Ministérios- Bloco D Anexo A Sala 330, Brasília, DF 70043-900, Brasil
FERREIRA DE SOUZA Marlene, Mapa, Esplanada dos Ministérios, Bloco D 8, Andar 830, Brasilia, DF 70043-900, Brasil
FERREIRA DUARTE Ceia, Fiscal Fed. Agropecuário, Lanagro/GO/CGAL/SDA/Mapa, Rua da Divisa s/n Setor Jaó, Goiânia, GO 74674-025, Brasil
FERREIRA Rachel, Mapa, Shin, Conj. 11 Casa 14, Lago Norte, Brasilia, DF 71515-110, Brasil
FIGUEIREDO DE CASTRO NASSAR Alessandra, Instituto Biológico, Av. Conselheiro Rodrigues Alves, 1252, São Paulo, SP 04014-002, Brasil

FONSECA JÚNIOR Antônio Augusto, Mapa, R. Rômulo Joviano, s/n, Caixa Postal 50, Pedro Leopoldo, MG 33.600-000, Brasil
FRANCO FIGUEIREDO Carlos Rogério, Mapa, SQN 106 Bloco A Apartamento 106, Asa Norte, Brasília, DF 70742-010, Brasil
FREIRE MARQUES Maria Risoleta, UFSC, Laboratorio de Biomarcadores de Contaminação Aquatica, Dep Bioquimica/Centro de Biociencias, Florianópolis, SC 88040-900, Brasil
FURTINI Ronaldo, IMA, Rua Maria Felipe de Araújo nº77, Apto 402 Bloco 2, Bairro Santa Efigênia, Belo Horizonte, MG 30270-470, Brasil

GARCIA Jose Fernando, UNESP, Aracatuba, Rua Clovis Pestana, 793 Jd D Amelia, Aracatuba, SP 16050-680, Brasil
GARDINER Christine, Canadian Food Inspection Agency, 3851 Fallowfield Rd., Fallowfield Rd., Ottawa, Ont. K2H 8P9, Canada
GENOVEZ,Margareth Elide, Instituto Biológico, Av. Cons. Rodrigues Alves 1252, São Paulo, SP 04014-002, Brasil
GEORGE Marcos, Ladives, Rio Tapia, Tocumen, Panama
GOLSTEYN-THOMAS Betty, Canadian Food Inspection Agency, P.O. Box 640, TWP RD 9-1, Lethbridge, Alberta TIJ 3Z4, Canada
GÓMES Concepción, Subdirección General de Sanidad Animal, CTRA, Algete Km 8, ES-28110 Madrid, Spain
GOMES DE SOUZA Jamil, Dpto. de Saúde Animal, Mapa, QE 02 bloco D, Apto 312, Guará I, DF 71010-040, Brasil
GOMES GUIMARÃES Ivens, Universidade Estadual de Londrina, Rua Piauí 854, Apartamento 101, Londrina, PR 86020-320, Brasil
GONÇALVES PINTO DA ROCHA Ana Cristina, Mapa, Shin Qi 12 Conj 7 Casa 9, Brasilia, DF 70000-000, Brasil
GONÇALVES Vitor Salvador P., UNB, Condomínio Villa de Montagne, Quadra 08 Casa 23, Brasília, DF 71.680-357, Brasil
GONGORA Victor, Animal Health, Belize Agricultural Health Authority, Mayavista, s/n, San Ignacio, Belize
GRANGER Larry M., USDA, Centers for Epidemiology and Animal Health, 2150 Centre Ave, Bldg B, Fort Collins, CO 80526, USA
GREGORI Fabio, Instituto Biológico, Laboratório de Doenças de Suínos, Av. Cons. Rodrigues Alves, 1252, Sâo Paulo, SP 014002, Brasil
GREISER-WILKE Irene, Virology, Buenteweg 17, DE-30559 Hannover, Germany
GRUZDEV Konstantin, FGI Arriah, Yurevets, RU-600901 Vladimir, Russia
GUIMARÃES Euripedes B., UFMS, Rua Tinhorão, Bairro Cidade Jardim, Campo Grande, MS 6796047-646, Brasil

HAKAN Vigre, Danish Institute for Food and Veterinary Research, Bulowsvej 27, DK-1790 Copenhagen V, Denmark
HANLON Cathleen, Centres for Disease Control, 1600 Clifton Rd, NE, Atlanta, GA 30333, USA
HERNANDES Maria Francisca, Ministério da Agricultura, Av. Rômulo Joviano S/N, Fazenda Modelo, Pedro Leopoldo, MG 33600-000, Brasil

JUNG Sukchan, National Veterinary Research & Quarantine Service, 480 Anyang 6-dong, Anyang-si, Kyounggi-do, 430-824 Anyang, Korea

KAZANTZI FONSECA André Salvador, Simbios Biotecnologia, Av. Farroupilha 8001, Prédio 22, 3º Andar, Canoas, RS 92425-900, Brasil
NISHIDA Marcelo Kenji, FMVZ/USP, Rua Doutor Gentil Leite Martin, No.242, Apt 71B, São Paulo, SP 04648-001, Brasil
KOENIG Patricia, Friedrich-Loeffler Institut/FLI, Boddenblick 5a, DE-17493 Greifswald-Insel Riems, Germany
KROTSCH Tomas Pedro, Instituto Interamericano de Cooperação para a Agri, Shis QI 03 Bloco F, Centro Empresarial Terracotta, Brasília DF 71605-450, Brasil
KUMAR SIVAM Shanta, VLA Weybridge, Woodham Lane, New Haw, Addlestone, Surrey KT15 3NB, U.K.

LAMBERT Catherine, AFSSA Fougères, La Haute Marche, FR-35300 Javène, France
LAUTNER Elizabeth A., National Veterinary Services Laboratories, 1800 Dayton Ave, Ames, IA 50010, USA
LOPES BRASIL Zelita de Oliveira, Lanagro/GO/CGAL/SDA/Mapa, Rua da Divisa S/N Setor Jaó, Goiânia, GO 74674-025, Brasil
LOURENÇO DA SILVA Paulo, Ufu, Rua João Severiano Rodrigues Da Cunha Nº 39, Jardim Caraíba, Uberlândia, MG 38411-178, Brasil
LOUVET CORTADA Véronique Micheline, Iagro, R. Nelson Figueiredo Junior 103 Casa 11, Chacara Vendas, Campo Grande, MS, 79003-210, Brasil
LOZA RUBIO Elizabeth, CENID, Carretera Hexicotoluca, Mexico City 5110, Mexico

MACEDO DE OLIVEIRA Anapolino, Lanagro/MG, av. Rômulo Joviano s/n faz. Modelo, Cx Postal 50, Pedro Leopoldo, MG 33600-000, Brasil
MADRUGA Claudio R., Embrapa Gado de Corte, Embrapa Gado de Corte Br, 262 km 4, Cx Postal 154, Compo Grande, MS 79002-970, Brasil
MAHANTACHAISAKUL Chaisiri, National Bureau of Agricultural Commodity and Food, 35/11 Soi Petchburi 5, Rachathewee Distric, Bangkok 10400, Thailand
MALDONADO ZAMBRANO Jesús A., Universidad Centroccidental Lisandro Alvarado, Parque Residencial La Mora, Conjunto 407, Torre D, Apto D12, La Mora, 3023 Cabudare, Estado Lara, Venezuela
MALIRAT Viviana, Centro Panamericano de Fiebre Aftosa OPS/OMS, R. Raul Pompéia, 141, Apto 103, Copacabana, Rio de Janeiro, RJ 22080-000, Brasil
MANFREDINI Mariangela, Idexx, Estrada das Rainhas 1800, Louveira, SP 13290-000, Brasil
MANNATHOKO Martin, Botswana Vaccine Institute, P/ Bag 0031, 31 Caborone, Botswana
MARREDO CONSTANTINOPOLO Soraya Elias, Lanagro/RS, Estrada Ponta Grossa, 3036, Porto Alegre, RS 91780-580, Brasil
MARTINEZ GÓMEZ Birmania, Comite Interamericano de Sanidad de Los Animales Acuáticos para Las Américas, OIE, Ministerio Agropecuario y Forestal, Magfor, Col 14 de Sept. I-678, 505 Managua, Nicaragua
MARTINS PINHEL Maria de Fatima, Lanagro/SP, Rua Raul Ferrari S/N, Campinas, SP 13100-105, Brasil
MATEJUSOVA Iveta, Fisheries Research Services, FRS Marine Laboratory, Aberdeen, AB11 9DB, U.K.
MATHIAS Luis Antonio, Unesp/Jaboticabal, Dep. de Medicina Vetrinaria Preventiva, Unesp, Jaboticabal, SP 14884-900, Brasil
MATOS CHAVES Humberto H., Unilan-Seagry, Av. Rui Barbosa Nº 1615, Aldeota Bairro, Fortaleza, CE 60410-300, Brasil
McGIVEN John A., VLA Weybridge, Woodham Lane, New Haw, Addlestone, Surrey, KT15 3NB, U.K.
MENDES RIBEIRO Simone Alves, Mapa, Rua Raul Ferrari S/N, Campinas, SP 13100-105, Brasil
MIKAMI Flordivina, Pecuária e Abasteciment, Sqn 105 Bloco C Aptº 404, Asa Norte, Brasília, DF 70734-030, Brasil
MOLITERNO DUARTE Dalila A., Mapa/Lanagro/PE, Rua Dom Manoel de Medeiros S/N, Bairro Dois Irmãos, Recife PE 52171-030, Brasil
MORAES ANDRADE Claudio, Pesagro/RJ, Rua Honório de Barros Nº 28, Flamengo, Rio de Janeiro, RJ 22250-120, Brasil
MORES Nelson, Centro Nacional de Pesquisa de Suino e Aves, Distrito de Tamanduá, Km 110, BR, 153 Cx. P. 21, Concordia, SC 89700-000, Brasil
MORGADO José C., Merial Saúde Animal Ltda, Av. Carlos Grimaldi, 1701, 4. andar, Campinas, SP 13091-908, Brasil
MORI ETIRO Ademar, Iagro, av. Filinto Mulher 146, Bairro Ipiranga, Campo Grande, MS 79 074-902, Brasil
MOZOLA Lindani, Quality Control and OIE Reference Laboratory, Botswana Vaccine Institute, Botswana, Private Bag 0031, 31 Gaborone, Botswana

NERY FARIAS DA SILVA Marta Maria, Lanagro/PE, Rua Dom Manoel de Medeiros S/N, Bairro Dois Irmãos, Recife, PE 52171-030, Brasil
NISHIZUMI Luciana Yumi, Mapa/SDA/CGAL/Lanagro/PA, Av. Almirante Borroso, 1234, Belem, PA 66095-000, Brasil

OLIVEIRA CAVALCANTI Luciaurea, Inmetro, Rua das Laranjeiras nº 347, Aptº 603, Laranjeiras, Rio de Janeiro, RJ 22240-002, Brasil
OLIVEIRA CUNHA Maria Izaura, Lanagro/PE, Rua Dom Manoel de Medeiros S/N, Recife, PE 52171-030, Brasil
O'NEIL Barry, Biosecurity New Zealand, Ministry of Agriculture and Forestry, PO Box 2526, 6015 Wellington, New Zealand
ORSI Maria Angela, Mapa ,Rua Raul Ferrari, S/N, Campinas, SP 13100-105, Brasil
ORTIZ RODRIGUEZ Nelly, Laboratory, Servicío Nacional de Calidad y Salud Animal, Ruta Mcal. Estigarribia Km 10,5, Não San Lorenzo, Paraguay
ORTIZ Tejedor, Subdirección General de Sanidad Animal, CTRA, Algete Km 8, ES-28110 Madrid, Spain

PADILHA DE ALENCAR Andréa, Lanagro/MG, av. Rômulo Joviano s/n Faz. Modelo, Cx Postal 50, Pedro Leopoldo, MG 33600-000, Brasil
PAULIN SILVA Lília Márcia, Instituto Biológico, Rua Padre Bento Ibáñez 545, Apto. 68, Bloco C, São Paulo, SP 04649-030, Brasil
PAZ RAMIREZ Pedro, Secretaria De Laboratorio, Alemania 16, 53260 Naucalpan, Mexico
PEPIN Michel, AFSSA, Alfort, 23, av. du General de Gaulle, FR-94706 Maisons-Alfort Cedex, France
PEREIRA Adriano, USDA/APHIS, Thomas Delorey, São Paulo, SP 04710-040, Brasil
PEREIRA Danilo, Governo do Estado de Santa Catarina, Rod Admar Gonzaga 1486, CP 1587, Florianopolis SC 88015-100, Brasil
PEREIRA MOURA Aitene, Embrapa Meio Norte, Br-343 K 35 Bairro Rural, Parnaiba, PI 64200-970, Brasil
PESSOA Gilson A., Universidade Federal de Santa Maria, CEU II Campus Universitário (UFSM), Apto 2321, Bairro Camobi, Av. Roraima, 1000, Santa Maria, RS 97105-900, Brasil
PINTO NASCIMENTO Ricardo Aurelio, Lanagro/MG, Av Romulo Joviano Sem Nº, Cx Postal 50, Pedro Leopoldo, MG 33600-000, Brasil
PIPPI SALLE Carlos Tadeu, UFRGS, av. Firmino Octávio Bimbi, 255/44ª, Bairro Cavalhada, Porto Alegre, RS 91751-330, Brasil
PITUCO Edviges M., Instituto Biológico, Rua Caravelas, 50, Ap. 82, São Paulo, SP 04012-060, Brasil
POTTS Lorraine Ann, Veterinary Laboratories Agency, Woodham Lane, New Haw, Addlestone, Surrey KT15 3NB, Brasil

RAVISON José A., Lanagro/RS, Paulo Derly Strehil, 128 Casa, Porto Alegre, RS 91771-640, Brasil
REGO PAMPLONA Ricardo, Pecuária e Abasteciment, Esplanada dos Ministérios Bloco D, Anexo A, Sala 447, Brasilia, DF 70043-900, Brasil
REISCHAK Dilmara, Mapa, Rua Raul Ferrari S/N, Campinas, SP 13100-105, Brasil
RENUNCIO BODANESE Eliana, Cooperativa Central Oeste Catarinense, Rua João Martins, 219 D, Chapecó, SC 89803-901, Brasil
RIBEIRO DE ARAUJO Flábio, Embrapa Gado de Corte, Embrapa Gado de Corte Br 262 Km 4, Caixa Postal 154, Compo Grande, MS 79002-970, Brasil
RIBEIRO DOS REIS Jorge, Mapa, Av. Almirante Barroso 1234, Belem, PA 66095-000 Brasil
RISTOW Luiz Eduardo, TECSA, Avenida do Contorno 6226, Belo Horizonte, MG 30110-110, Brasil
RIVERA SALAZAR Alejandro, Vigilancia Epidemiológica, Servicio Agrícola y Ganadero, Bulnes 140 Santiago, 8330-246 Santiago, Chile
RIZALDO PINHEIRO Raymundo, Embrapa Caprinos, Estr. Sobral/Groaíras, Km 04- Rural, Sobral, CE 62010-970, Brasil
ROMERO Igor, Servicio Nacional de Sanidad Agropecuaria, Municipio Libe 322, 6100 Mexico, Mexico
ROSENDO DO NASCIMENTO Elmiro, UFF, Rua Vital Brazil Filho 64, Vital Brazil, Niteroi, RJ 24230-340, Brasil
ROTH James, Center for Food Security and Public Health, 2156 College of Veterinary Medicine, Ames, IA 50010, USA
RUSSIANO Vicente Wilter Ricardo, FCAV, Unesp Jaboticabal, Rua Caravelas, 50 Ap. 82, Ribeirão Preto, Ribeirão Preto, São Paulo, SP 04012-060, Brasil

SAENZ ACHURRA Luis Arturo, Ministerio de Desarrollo Agropecuario, Río Tapia, Tocumen, 81601-611, Panamá.

SAGEL Agustin, Comisión Panamá, Segunda Entrada, David, Panama

SALVATIERRA OPORTO Jaime, CIDASC, Joaquim Vaz 1661, São José, SC 86103-215, Brasil

SANTIAGO SILVA Virginia, Embrapa, Br 153 Km 110 Distrito de Tamanduá, Cx. P. 21, Concordia, SC 89700-000, Brasil

SATOMI KUCHIISHI Suzana, Cedisa, BR 153 Km 110, Vila Tamanduá, Concórdia, SC 89700-000, Brasil

SAUCEDO ALDERETE Raul, Auturara Panamena de Seguridad de Alimento, Sun Towor Plaza Piso 1, Panama, Panama

SCANDRETT Brad, Canadian Food Inspection Agency, Saskatoon, Saskatchewan S7N 2R3, Canada

SCHAEFER Rejane, Centro Nacional de Pesquisa de Suino, Distrito de Tamanduá, Km 110, BR 153 Cx.P. 21, Concordia, SC 89700-000, Brasil

SEQUETIN CUNHA Elenice M., Instituto Biológico, Av. Cons. Rodrigues Alves, 1252, São Paulo, SP 04014-002, Brasil

SERRANO RAMIREZ Emerio, Instituto de Medicina Veterinaria, Ministerio de la Agricultura, 12 # 355 15 y 17, La Habana 14000, Cuba

SESTAK Leonardo, Cobb-vantress Brasil Ltda, Rua Gabriel Janikian, 280 Parque Quintas das Paineiras, São José do Rio Preto, SP15 080-350, Brasil

SESTI Luiz, Ceva Saúde Animal Ltda., Rua Santa Justina 660, Vila Olímpia, São Paulo, SP 04545-042, Brasil

SOARES FILHO Paulo Martins, Ministerio da Agricultura/Lanagro, Av. Rõmulo Joviano S/N, Centro Cx Postal 50, Pedro Leopoldo, MG 33600-000, Brasil

SOARES LEITE Adriana, Lanagro/Pe, Rua Dom Manoel de Medeiros S/N, Recife PE 52171-030, Brasil

SOLARI Claude-André, Rua Soares Cabral Nº 80 Aptº 1204, Laranjeiras, Rio de Janeiro RJ 22240-070, Brasil

SONCINI Ricardo Alfredo, Sadia S.A., Rua Senador Attilio Fontana # 86, Bairro Centro, Concordia, SC 89700-000, Brasil

SONNTAG KUCHENBECKER Beatris, Lanagro/RS, Av. Mathilde Trein Tenner, 251, Apt 601, Porto Alegre, RS 91370-050, Brasil

SPACAGNA JORDÃO Ricardo, Instituto Biológico, Rua Visconde de Inhaúma, 480, Apto 32-B, São Paulo, SP 4145-030, Brasil

STAHL Karl, SVA, SE-75189 Uppsala, Sweden

TADEU BARBOSA Roni, Secretaria de Agricultura de Santa Catarina, Rodovia Admar Gonzaga,1486, Florianopolis, SC 88034-001, Brasil

TAYLOR Andrew Vaughan, VLA Weybridge, Woodham Lane, New Haw, Addlestone, Surrey KT15 3NB, U.K.

THOBOKWE Gaolathe, Botswana Vaccine Institute, Private Bag 0031, 31 Gaborone, Botswana

TOSCHI MACIEL Maria Luiza, Ministério da Agricultura, R: Felipe Schmith, 755, Sala 802, Florianópolis, SC 88010-002, Brasil

TREVISOL Iara, Centro Nacional de Pesquisa de Suino, Distrito de Tamanduá, Km 110 BR 153, Cx Postal 21, Concordia, SC 89700-000, Brasil

TURILLI Carlo, Uniceb, Rua Joao Wyclif, 255, Apt 1203, Londrina, PR 86050-450, Brasil

VACAS DE CARVALHO CORREA DE SA Maria Inácia, Lab. Nac. Inv. Veterinária, Estrada de Benfica, 701, PT-1549-011 Lisboa, Portugal

VASCONCELLOS CARDOSO Maristela, Pesquisador Científico, Instituto Biológico, Av. Cons. Rodrigues Alves, 1252, CPDSA, São Paulo, SP 04014-002, Brasil

VASCONCELOS Anilton Cesar, UFMG, Rua Carangola Nº 57 Apto 1202, Bairro Santo Antonio, Belo Horizonte, MG 30330-240, Brasil

VASCONCELOS DE OLIVEIRA Cynthia A., Lanagro/PE, Rua João Ramos, 171/904 Graças, Recife, PE 52011-080, Brasil

VAZ DA COSTA BARBOSA Alessandra C., Lanagro/Go/Cgal/Sda/Mapa, Rua Da Divisa S/N Setor Jaó, Goiânia, GO 74674-025, Brasil

VENTURA Lauren, Cedisa, BR 153 Km 110, Vila Tamanduá, Concórdia, SC 89700-000, Brasil

VIANNA DE FREITAS João, Lanagro/RS, Mariz E Barros, 36, Apto 403, Porto Alegre, RS 90690-390, Brasil

VIEIRA BENAVIDES Magda, Embrapa, Zona Rural, Bagé, RS 96401-970, Brasil
VIEIRA SERRA Claúdia, Lanagro/MG, Av. Rômulo Joviano s/n Faz. Modelo, Cx Postal 50, Pedro Leopoldo, MG 33600-000, Brasil
VILELA VIEIRA Fernando, ABS Pecplan, Rod. BR 050 Km 195, CP 4046, Uberaba, MG 38020-300, Brasil
VINÍCIUS DE S. LEANDRO JUNIOR Marcos, Ministério da Agricultura, Pecuária e Abasteci ment, SQS 208, Bloco K Apt° 104, Asa Sul, Brasília, DF 70254-110 Brasil
VITOR KOERICH HERN Priscilla Karina, Perdigao, Rodovia SC, 453 Km 50, Videira, SC 89560-000, Brasil

WANDERLEY DA NOBREGA LIRA Seleucia, SFA/AC, Conjunto Tropical I Qd B Casa 35, Bairro São Francisco, Rio Branco, AC 69910-490, Brasil

ZANATTA STOPPA Greice F., Inst. Biológico, Rua Cel Rafael Tobias 184, Descalvado, SP 13690-000, Brasil
ZMUDZINSKI Jan F., National Veterinary Res. Institute, Al. Partyzantow 57, PL-24-100 Pulawy, Poland

Index of Authors

Developments in Biological Standardization

Vol. 36	Washington DC, 1976	Freeze-Drying of Biological Products
Vol. 37	Geneva, 1976	Standardization of Cell Substrates for the Production of Virus Vaccines
Vol. 38	London, 1977	Biological Preparations in the Treatment of Cancer
Vol. 39	Geneva, 1977	Influenza Immunization
Vol. 40	Marburg, 1977	Standardization of Rabies Vaccines for Human Use Produced in Tissue Culture (Rabies III)
Vol. 41	Guadeloupe, 1978	Vaccines in the Developing Countries
Vol. 42	Paris, 1978	Second General Meeting of ESACT
Vol. 43	Brussels, 1978	Immunization: Benefit Versus Risk Factors
Vol. 44	Villars, 1979	Test methods for the Quality Control of Plasma Proteins
Vol. 45	San Antonio, 1979	The Standardization of Animals to Improve Biomedical Research, Pro duction and Control
Vol. 46	Oxford, 1979	Third General Meeting of ESACT
Vol. 47	Bilthoven, 1980	Reassessment of Inactivated Poliomyelitis Vaccine
Vol. 48	Geneva, 1980	Standardization of Albumin, Plasma Substitutes and Plasmapheresis
Vol. 49	Leetown, 1981	Fish Biologics: Serodiagnostics and Vaccines
Vol. 50	Heidelberg, 1981	The Use of Heteroploid and Other Cell Substrates for the Production of Biologicals
Vol. 51	Lyons, 1981	Immunization of Adult Birds with Inactivated Oil Adjuvant Vaccines
Vol. 52	Lyons, 1981	Herpes Virus of Man and Animal: Standardization of Immunological Procedures
Vol. 53	Dublin, 1982	Enteric Infections in Man and Animals: Standardization of Immunologi cal Procedures
Vol. 54	Athens, 1982	Viral Hepatitis: Standardization in Immunoprophylaxis of Infections by Hepatitis Viruses
Vol. 55	Copenhagen, 1982	Fifth General Meeting of ESCAT
Vol. 56	Algiers, 1983	Third International Symposium on Brucellosis
Vol. 57	Paris, 1983	Monoclonal Antibodies: Standardization of their Characterization and Use
Vol. 58	Budapest, 1983	International Symposium on BCG Vaccines and Tuberculins
Vol. 59	Geneva, 1983	Standardization and Control of Biologicals Produced by Recombinant DNA Technology
Vol. 60	Gardone Riviera, 1984	Production and Exploitation of Existing and New Animal Cell Sub strates
Vol. 61	Geneva, 1984	Proceedings of the Fourth International Symposium on Pertussis
Vol. 62	Stockholm, 1985	Diagnostics and Vaccines for Parasitic Diseases
Vol. 63	San Francisco, 1985	Use and Standardization of Chemically Defined Antigens
Vol. 64	London, 1985	Reduction of Animal Usage in the Development and Control of Biologi cal Products
Vol. 65	Amsterdam, 1985	Use and Standardization of Combined Vaccines
Vol. 66	Vienna, 1985	Advances in Animal Cell Technology: Cell Engineering, Evaluation and Exploitation
Vol. 67	Melbourne, 1986	Standardization in Blood Fractionation Including Coagulation Factors
Vol. 68	Geneva, 1986	Cells - Products - Safety
Vol. 69	London, 1987	Cytokines Laboratory and Clinical Evaluation
Vol. 70	Arlington, VA, 1988	Continuous Cell Lines as Substrates for Biologicals

Vol. 71	Utrecht, 1989	Monoclonal Antibodies for Therapy, Prevention and In vivo Diagnosis of Human Disease
Vol. 72	Annecy, 1989	Progress in Animal Retroviruses
Vol. 73	Shizuoka, 1990	Pertussis: Evaluation and Research on Acellular Pertussis Vaccines
Vol. 74	Bethesda, MD, 1990	Biological Product Freeze-Drying and Formulation
Vol. 75	London, 1990	Virological Aspects of the Safety of Biological Products
Vol. 76	Bethesda, MD, 1991	Continuous Cell Lines - An International Workshop on Current Issues
Vol. 77	Annecy, 1991	Standardization of the Immunopharmacology of Natural and Synthetic Immunomodulators
Vol. 78	Bethesda, MD, 1991	Poliovirus Attenuation: Molecular Mechanisms and Practical Aspects
Vol. 79	Ploufragan, 1992	The First Steps Towards an International Harmonization of Veterinary Biologicals: 1993 and Free Circulation of Vaccines within the EEC
Vol. 80	Heidelberg, 1992	Transmissible Spongiform Encephalopathies - Impact on Animal and Human Health
Vol. 81	Cannes, 1992	Virological Safety Aspects of Plasma Derivatives
Vol. 82	Albany NY, 1993	Recombinant Vectors in Vaccine Development
Vol. 83	Annecy, 1993	Genetic Stability and Recombinant Product Consistency
Vol. 84	Langen, 1993	Non-Target Effects of Live Vaccines
Vol. 85	Santa Fé, NM, 1994	Genetics of Streptococci, Enterococci and Lactococci
Vol. 86	Langen, 1994	Replacement, Reduction and Refinement of Animal Experiments in the Development and Control of Biological Products
Vol. 87	Geneva, 1995	New Approaches to Stabilisation of Vaccines Potency
Vol. 88	Bethesda, MD, 1995	Viral Safety and Evaluation of Viral Clearance from Biopharmaceutical Products
Vol. 89	Rome, 1995	Pertussis Vaccine Trials
Vol. 90	Oslo, 1996	Fish Vaccinology
Vol. 91	San Francisco, CA, 1996	Development of Specifications for Biotechnology Pharmaceuticals Products
Vol. 92	Bergen, 1996	Modulation of the Immune Response to Vaccine Antigens
Vol. 93	Annecy, 1996	Safety of Biological Products Prepared from Mammalian Cell Culture
Vol. 94	Bethesda, MD, 1997	Simian Virus 40 (SV40): A Possible Human Polyomavirus
Vol. 95	Paris, 1997	Preclinical and Clinical Development of New Vaccines
Vol. 96	Washington, DC, 1995	Characterization of Biotechnology Pharmaceutical Products
Vol. 97	London, 1997	Biological Characterization and Assay of Cytokines and Growth Factors
Vol. 98	London, 1997	Inactivated Influenza Vaccines Prepared in Cell Culture
Vol. 99	Strasbourg, 1998	Animal Sera, Animal Sera Derivatives and Substitutes Used in the Man ufacture of Pharmaceuticals: Viral Safety and Regulatory Aspects
Vol. 100	Geneva, 1998	A Celebration of 50 Years of Progress in Biological Standardization and Control at WHO
Vol. 101	London, 1998	Alternatives to Animals in the Development and Control of Biological Products for Human and Veterinary Use

Developments in Biologicals

Vol. 102	San Francisco, CA, 1999	Advances in Transfusion Safety
Vol. 103	Annecy, 1999	Physico-Chemical Procedures for the Characterization of Vaccines
Vol. 104	Langen, 1999	Development and Clinical Progress of DNA Vaccines
Vol. 105	Paris, 2000	Progress in Polio Eradication: Vaccine Strategies for the End Game
Vol. 106	Rockville, MD, 2000	Evolving Scientific and Regulatory Perspectives on Cell Substrates for Vaccine Development
Vol. 107	London, 2000	The Design and Analysis of Potency Assays for Biotechnology Products
Vol. 108	Langen, 2001	Advances in Transfusion Safety
Vol. 109	Washington, DC, 2000	Biologics 2000 – Comparability of Biotechnology Products
Vol. 110	Palm Cove, 2001	Orphan Vaccines – Bridging the Gap
Vol. 111	Utrecht, 2001	Advancing Science and Elimination of the Use of Laboratory Animals for Development and Control of Vaccines and Hormones
Vol. 112	Bethesda, MD, 2001	Immunogenicity of Therapeutic Biological Products
Vol. 113	Berlin, 2001	Process Validation for Manufacturing of Biologics and Biotechnology Products
Vol. 114	Ames, IA, 2002	Vaccines for OIE List A and Emerging Animal Diseases
Vol. 115	Bergen, 2002	Laboratory Correlates of Immunity to Influenza – A Reassessment,
Vol. 116	Los Angeles, CA, 2003	Development of Therapeutic Cancer Vaccines
Vol. 117	Langen, 2003	Consideration of Alternative Licensing Procedures for Vaccines for Minor Species, Minor Indications and Autogenous/Autologous Products
Vol. 118	Langen, 2003	PDA/EMEA European Virus Safety Forum
Vol. 119	Buenos Aires, 2004	Control of Infectious Animal Diseases by Vaccination
Vol. 120	Bethesda, MD, 2003	Advances in Transfusion Safety
Vol. 121	Bergen, 2003	Progress in Fish Vaccinology
Vol. 122	Bethesda, MD, 2003	State of the Art Analytical Methods for Characterization of Biological Products and Assessment of Comparability
Vol. 123	Rockville, MD, 2004	Vaccine Cell Substrates 2004
Vol. 124	Paris, France, 2005	OIE/FAO International Scientific Conference on Avian Influenza
Vol. 125	Kiev, Ukraine, 2005	First International Conference on Rabies in Europe
Vol. 126	Saint-Malo, France, 2005	New Diagnostic Technology: Applications in Animal Health and Biologics Controls
Vol. 127	Sydney, Australia, 2005	Advances in Transfusion Safety – Volume IV

IMPRIMERIE MEDECINE M H HYGIENE GENEVE SUISSE

septembre–2007